Mitsuo Gen, David Green, Osamu Katai, Bob McKay, Akira Namatame, Ruhul A. Sarker, and Byoung-Tak Zhang (Eds.)

# Intelligent and Evolutionary Systems

# Studies in Computational Intelligence, Volume 187

**Editor-in-Chief**
Prof. Janusz Kacprzyk
Systems Research Institute
Polish Academy of Sciences
ul. Newelska 6
01-447 Warsaw
Poland
*E-mail:* kacprzyk@ibspan.waw.pl

---

Further volumes of this series can be found on our homepage: springer.com

Vol. 165. Djamel A. Zighed, Shusaku Tsumoto, Zbigniew W. Ras and Hakim Hacid (Eds.)
*Mining Complex Data*, 2009
ISBN 978-3-540-88066-0

Vol. 166. Constantinos Koutsojannis and Spiros Sirmakessis (Eds.)
*Tools and Applications with Artificial Intelligence*, 2009
ISBN 978-3-540-88068-4

Vol. 167. Ngoc Thanh Nguyen and Lakhmi C. Jain (Eds.)
*Intelligent Agents in the Evolution of Web and Applications*, 2009
ISBN 978-3-540-88070-7

Vol. 168. Andreas Tolk and Lakhmi C. Jain (Eds.)
*Complex Systems in Knowledge-based Environments: Theory, Models and Applications*, 2009
ISBN 978-3-540-88074-5

Vol. 169. Nadia Nedjah, Luiza de Macedo Mourelle and Janusz Kacprzyk (Eds.)
*Innovative Applications in Data Mining*, 2009
ISBN 978-3-540-88044-8

Vol. 170. Lakhmi C. Jain and Ngoc Thanh Nguyen (Eds.)
*Knowledge Processing and Decision Making in Agent-Based Systems*, 2009
ISBN 978-3-540-88048-6

Vol. 171. Chi-Keong Goh, Yew-Soon Ong and Kay Chen Tan (Eds.)
*Multi-Objective Memetic Algorithms*, 2009
ISBN 978-3-540-88050-9

Vol. 172. I-Hsien Ting and Hui-Ju Wu (Eds.)
*Web Mining Applications in E-Commerce and E-Services*, 2009
ISBN 978-3-540-88080-6

Vol. 173. Tobias Grosche
*Computational Intelligence in Integrated Airline Scheduling*, 2009
ISBN 978-3-540-89886-3

Vol. 174. Ajith Abraham, Rafael Falcón and Rafael Bello (Eds.)
*Rough Set Theory: A True Landmark in Data Analysis*, 2009
ISBN 978-3-540-89886-3

Vol. 175. Godfrey C. Onwubolu and Donald Davendra (Eds.)
*Differential Evolution: A Handbook for Global Permutation-Based Combinatorial Optimization*, 2009
ISBN 978-3-540-92150-9

Vol. 176. Beniamino Murgante, Giuseppe Borruso and Alessandra Lapucci (Eds.)
*Geocomputation and Urban Planning*, 2009
ISBN 978-3-540-89929-7

Vol. 177. Dikai Liu, Lingfeng Wang and Kay Chen Tan (Eds.)
*Design and Control of Intelligent Robotic Systems*, 2009
ISBN 978-3-540-89932-7

Vol. 178. Swagatam Das, Ajith Abraham and Amit Konar
*Metaheuristic Clustering*, 2009
ISBN 978-3-540-92172-1

Vol. 179. Mircea Gh. Negoita and Sorin Hintea
*Bio-Inspired Technologies for the Hardware of Adaptive Systems*, 2009
ISBN 978-3-540-76994-1

Vol. 180. Wojciech Mitkowski and Janusz Kacprzyk (Eds.)
*Modelling Dynamics in Processes and Systems*, 2009
ISBN 978-3-540-92202-5

Vol. 181. Georgios Miaoulis and Dimitri Plemenos (Eds.)
*Intelligent Scene Modelling Information Systems*, 2009
ISBN 978-3-540-92901-7

Vol. 182. Andrzej Bargiela and Witold Pedrycz (Eds.)
*Human-Centric Information Processing Through Granular Modelling*, 2009
ISBN 978-3-540-92915-4

Vol. 183. Marco A.C. Pacheco and Marley M.B.R. Vellasco (Eds.)
*Intelligent Systems in Oil Field Development under Uncertainty*, 2009
ISBN 978-3-540-92999-4

Vol. 184. Ljupco Kocarev, Zbigniew Galias and Shiguo Lian (Eds.)
*Intelligent Computing Based on Chaos*, 2009
ISBN 978-3-540-95971-7

Vol. 185. Anthony Brabazon and Michael O'Neill (Eds.)
*Natural Computing in Computational Finance*, 2009
ISBN 978-3-540-95973-1

Vol. 186. Chi-Keong Goh and Kay Chen Tan
*Evolutionary Multi-objective Optimization in Uncertain Environments*, 2009
ISBN 978-3-540-95975-5

Vol. 187. Mitsuo Gen, David Green, Osamu Katai, Bob McKay, Akira Namatame, Ruhul A. Sarker and Byoung-Tak Zhang (Eds.)
*Intelligent and Evolutionary Systems*, 2009
ISBN 978-3-540-95977-9

Mitsuo Gen
David Green
Osamu Katai
Bob McKay
Akira Namatame
Ruhul A. Sarker
Byoung-Tak Zhang
(Eds.)

# Intelligent and Evolutionary Systems

Springer

Mitsuo Gen
Waseda University
Graduate School of IPS
2-8 Hibikino
Wakamatsu-ku, Kitakyushu 808-0135
Japan
E-mail: gen@waseda.jp

David Green
Clayton School of Information Technology
Monash University
Clayton Victoria 3800, Australia
E-mail: david.green@infotech.monash.edu.au

Osamu Katai
Dept. of Systems Science
Graduate School of Informatics
Kyoto University
Sakyo-ku, Kyoto 606-8501, Japan
E-mail: katai@i.kyoto-u.ac.jp

Bob McKay
School of Computer Science and Engineering
Seoul National University
Gwanangno 599
Seoul 151-744, Korea
E-mail: rim@cse.snu.ac.kr

Akira Namatame
Dept. of Computer Science
National Defense Academy of Japan
Yokosuka, 239-8686, Japan
E-mail: nama@nda.ac.jp

Ruhul Sarker
School of IT&EE, UNSW@ADFA
Northcott Dve
Campbell, ACT2600, Australia
E-mail: r.sarker@adfa.edu.au

Byoung-Tak Zhang
School of Computer Science and Engineering
Seoul National University
Gwanangno 599
Seoul 151-744, Korea
E-mail: btzhang@bi.snu.ac.kr

ISBN 978-3-642-10114-4 e-ISBN 978-3-540-95978-6

DOI 10.1007/978-3-540-95978-6

Studies in Computational Intelligence ISSN 1860949X



*Typeset & Cover Design:* Scientific Publishing Services Pvt. Ltd., Chennai, India.

Printed in acid-free paper

9 8 7 6 5 4 3 2 1

springer.com

# Preface

Artificial evolutionary systems are computer systems, inspired by ideas from natural evolution and related phenomena. The field has a long history, dating back to the earliest days of computer science, but it has only become an established scientific and engineering discipline since the 1990s, with packages for the commonest form, genetic algorithms, now widely available.

Researchers in the Asia-Pacific region have participated strongly in the development of evolutionary systems, with a particular emphasis on the evolution of intelligent solutions to highly complex problems. The Asia-Pacific Symposia on Intelligent and Evolutionary Systems have been an important contributor to this growth in impact, since 1997 providing an annual forum for exchange and dissemination of ideas. Participants come primarily from East Asia and the Western Pacific, but contributions are welcomed from around the World.

This volume features a selection of fourteen of the best papers from recent APSIES. They illustrate the breadth of research in the region, with applications ranging from business to medicine, from network optimization to the promotion of innovation.

It opens with three papers in the general area of business and economics. Orito and colleagues extend previous work on the application of evolutionary algorithms to index fund optimization by incorporating local search in an unusual way: using the genetic search to maximize the coefficient of determination between the fund's return rate and the market index (but not necessarily finding a linear relationship), and then using local search to optimize the linearity. They demonstrate that this approach outperforms direct search, yielding funds that perform substantially better as a surrogate for the Tokyo Stock Price Index from 1997 to 2005.

Guo and Wong investigate the problem of learning Bayesian Networks from incomplete data. They modify their group's previous hybrid evolutionary algorithm for learning from complete data. It uses essentially Friedman's Structural Expectation Maximization (SEM) algorithm as the outer loop, with a variant of their evolutionary algorithm in the inner loop, replacing SEM's hill-climbing phase. It differs from previous algorithms, which use the expected value to replace missing values, in using a more sophisticated data completion process, which permits the use of decomposable scoring metrics (specifically, information-based metrics) in the search process. They use the algorithm in a direct-marketing application, demonstrating improved performance on that problem, though the technique would clearly extend to other domains – DNA chip analysis, ecological data – where missing values cause serious difficulties.

Katai and his colleagues consider cooperative or 'local' currencies, and investigate the design of such currencies to promote social and economic goals. They base their analysis on fuzzy theory, and obtain interesting new results on the desirable operation of such systems.

Networks have become a key area of complex systems research, with applications ranging from communications to transport problems to the organisation of web pages. The next six papers exemplify this trend, examining various aspects of network theory.

Leu and Namatame consider the problem of failure resilience in networks, such as power distribution or communications networks. They apply evolutionary algorithms to optimising the robustness of such networks to link failure, and are able to demonstrate that, under certain circumstances, they are able to preserve important linkage properties of the networks (notably, scale-freeness), while improving the failure resilience.

While Leu and Namatame consider robustness to link breakages in networks, Newth and Ash consider instead robustness to disturbance, and the linearity of network response. Again, they apply an evolutionary algorithm to optimise robustness. They observe an interesting property, that the optimised networks they evolve exhibit hub-and-star like topology, suggesting that this structure has inherent stability properties.

Komatsu and Namatame propose a heterogeneous flow control mechanism for protecting communications networks from attacks such as DDoS. They distinguish between altruistic protocols such as tcp, and uncontrolled protocols such as udp, using open-loop congestion control mechanisms such as drop-tail for the former, and closed-loop such as RED and CHOKe for the latter. Using simulations on a range of network topologies, they demonstrate good performance in controlling excess traffic by com parison with homogeneous protocols, and propose extensions of this approach to higher layers in the protocol stack.

Lin and Gen concentrate on the problem of network routing, specifically on finding Shortest Path Routes (SPR) for Open Shortest Path First (OSPF) routing protocols. They propose a new priority-based representation and genetic algorithm for this problem, and demonstrate its performance through a range of numerical experiments.

Network flow problems are a classic problem in the optimization literature; Gen, Lin and Jo extend the usual problem, of maximizing network flow, into a bi-criteria problem, maximizing network flow while minimizing network cost. They report on a variant evolutionary multi-objective optimization algorithm incorporating Lamarckian local search, and demonstrate its performance on a range of test problems.

A second paper from the same authors considers applications in logistics network design, starting from the design of the network, and extending to vehicle routing and automated vehicle dispatch. They introduce a priority-based Genetic Algorithm for the task, applying variants to all three problems, with good results.

The final paper on network problems, by Lin and Gen, approaches the problem of bi-criteria design of networks from a more general perspective. To illustrate their approach, they tackle three separate design problems:

1. Shortest path, in which the conflicting objectives are to minimize transmission delay while at the same time minimizing network cost
2. Spanning tree, in which the conflicting objectives are as above (i.e. minimizing both transmission delay and network cost)

3. Network flow, in which the conflicting objectives are to maximize network flow while at the same time minimizing network cost

The authors compare a number of representations and algorithms for these problems, generating interesting results showing that complex versions of these problems can realistically be solved with today's algorithms.

Sawazumi et al. investigate mechanisms to promote human creativity, proposing a method based on "serendipity cards", cards containing detailed information about a theme. In so doing, they introduce a number of ideas and contexts from the Japanese literature on idea generation not well known outside of Japan.

Cornforth et al tackle an important medical problem, that of recognition of medical problems from imagery. Specifically, they concentrate on the issue of medical image segmentation, in the context of assessment of retinopathy due to diabetes. They combined wavelet data extraction methods with Gaussian mixture Bayesian classifiers, generating substantially improvements over simpler methods, though not quite matching expert-level human performance.

Gen et al tackle another highly practical problem, the problem of job-shop scheduling in a shop where some machines may substitute for others for particular operations (in the classical job-shop scheduling problem, each operation can be performed on precisely one machine). They introduce a new multi-stage genetic algorithm, comparing it with the state of the art in the field. They demonstrate very substantially improved performance over a classical genetic algorithm, and GA augmented with a form of local search, especially on hard problems. They demonstrate some improvement in comparison with a particle-swarm/simulated annealing hybrid method, though the differences are small.

Wong and Wong round out the volume with a paper showing that impressive speed of evolutionary algorithms may be obtained at relatively low cost, through implementation on graphics processing units. They obtain very impressive performance indeed on a range of benchmark optimization problems, especially for large population sizes.

Overall, the papers represent just a sample of the wide range of research in intelligent and evolutionary systems being conducted in the Asia- Pacific region.. The growing maturity of its research culture portends an increasing contribution to international research across the range of the sciences, and in intelligent systems in particular. We hope this volume can serve as a stepping stone in this process, introducing some of the work to a wider audience, and at the same time increasing international awareness of one of this Asia-Pacific forum.

November 2008

Mitsuo Gen
David Green
Osamu Katai
Bob McKay
Akira Namatame
Ruhul Sarker
Byoung-Tak Zhang

# Contents

# Index Fund Optimization Using Genetic Algorithm and Scatter Diagram Based on Coefficients of Determination

Yukiko Orito[1], Manabu Takeda[2], and Hisashi Yamamoto[2]

[1] Ashikaga Institute of Technology
268-1, Ohmae-cho, Ashikaga, Tochigi 326-8558, Japan
orito@ashitech.ac.jp

[2] Tokyo Metropolitan University
6-6, Asahigaoka, Hino, Tokyo 191-0065, Japan
u05851413@cc.tmit.ac.jp, yamamoto@cc.tmit.ac.jp

Index fund optimization is one of portfolio optimizations and can be viewed as a combinatorial optimization for portfolio managements. It is well known that an index fund consisting of stocks of listed companies on a stock market is very useful for hedge trading if the total return rate of a fund follows a similar path to the rate of change of a market index. In this paper, we propose a method that consists of a genetic algorithm and a heuristic local search on scatter diagrams to make linear association between the return rates and the rates of change strong. A coefficient of determination is adopted as a linear association measure of how the return rates follow the rates of change. We then apply the method to the Tokyo Stock Exchange. The results show that the method is effective for the index fund optimization.

**Keywords:** Index Fund Optimization; Coefficient of Determination; Genetic Algorithm; Heuristic Local Search.

## 1 Introduction

Index fund optimization is one of portfolio optimizations and can be viewed as a combinatorial optimization for portfolio managements. It is well known that a group consisting of stocks of listed companies on a stock market is very useful for hedge trading if the total return rates of a group follow a similar path to the rates of change of a market index. Such a group is called an index fund. An index fund has been used very extensively for the hedge trading, which is the practice of offsetting the price risk on any cash market position by taking an equal, but opposite position in a futures market [1]. In addition, there are some studies report that the index funds have better performance than other mutual funds [2, 3, 4].

The index fund optimization problem is one of NP-complete problems, and it is impossible to solve it in reasonable time when the number of listed

M. Gen et al.: Intelligent and Evolutionary Systems, SCI 187, pp. 1–11.
springerlink.com 

companies or the number of stocks of each company exceeds some not-so-very-large numbers. In order to solve this problem, most efforts are focused on finding the optimal solution through large space searching methods such as evolution algorithms.

In the portfolio optimization field, Xia et al. [5] provided optimal portfolios to achieve maximum return and minimum risk by using a Genetic Algorithm (GA). Chang et al. [6] compared a tabu search with a simulated annealing for portfolio optimizations. In the index fund optimization field, Oh et al. [7] showed the effectiveness of index funds optimized by a GA on the Korean Stock Exchange. Takabayashi [8] proposed a GA method to select listed companies on the Tokyo Stock Exchange. Although his GA method selected companies to an index fund, it did not optimize the proportion of funds in the index fund. On the other hand, index funds require rebalancing in order to reflect the changes in composition of the market index over the fund's future period. However, the total price of index fund is unknown, so the implied cost of rebalancing is uncertain. If we invest a lot of money in an index fund, we have to make a great investment in rebalancing (for discussion of rebalancing cost, see, e.g., Aiello and Chieffe [9] and Chang [10]). In this context, it is desired that the index fund consists of a small number of companies. Orito et al. [11] have proposed a GA method that optimized the proportion of funds in an index fund under cost constraints. Their GA method first chooses a certain number of companies on a market by using a heuristic rule and then applies a GA to optimize the proportion of funds for these chosen companies. Although their methods are superior to Takabayashi [8]'s method, the results of the numerical experiments did not show satisfactory efficiency.

In this paper, we propose a new method consisting of the following two steps:

**Step 1.** Optimize the proportion of funds in an index fund under cost constraints by using a GA. This step is based on the methodology of Orito et al. [11].
**Step 2.** Re-optimize the proportion of funds in the index fund by using a heuristic local search on scatter diagrams. This is the step that we newly propose in this paper.

This paper is structured as follows: Section 2 introduces the index fund optimization problem and formulates it. In Section 3, we propose a method to optimize the proportion of fund for an index fund. Section 4 demonstrates the effectiveness of our method through numerical experiments. We then conclude this research work in Section 5.

## 2 The Index Fund Optimization Problem

In this section, we describe the index fund optimization problem. First, we define the following notations.

$N$: the number of listed companies in a fund.
$i$: Company $i$, $i = 1, 2, \cdots, N$.

$g_i$: the proportion of funds for Company $i$.
$\bar{g}$: the fund (portfolio). That is, $N$-dimensional vector $\bar{g} = (g_1, g_2, \cdots, g_N)$ such that $\sum_{i=1}^{N} g_i = 1$.
$t$: time basis, dates data $t = 0, 1, \cdots, T$.
$Q(t)$: the market index at $t$.
$x(t)$: the rate of change of market index at $t$. That is $x(t) = \dfrac{Q(t+1) - Q(t)}{Q(t)}$.
$P_i(t)$: the stock price of Company $i$ at $t$.
$V_i(g_i; t)$: the present value of Company $i$ at $t$. This is defined by Equation (1).
$F(\bar{g}; t)$ : the total price of the fund $\bar{g}$ at $t$. That is $F(\bar{g}; t) = \sum_{i=1}^{N} V_i(g_i; t)$.
$y(\bar{g}; t)$: the return rate of the fund $\bar{g}$ at $t$. That is $y(\bar{g}; t) = \dfrac{F(\bar{g}; t+1) - F(\bar{g}; t)}{F(\bar{g}; t)}$.
$R^2(\bar{g})$: the coefficient of determination between the fund's return rates $y(\bar{g}; t)$ and the rates of change of market index $x(t)$ over $T$ dates data. This is defined by Equation (2).

Suppose that we invest in a fund consisting of $N$ listed companies, Companies $1, 2, \cdots, N$, which starts at $t = 0$ and ends at $t = T$. The present value for each Company $i$ at $t$ is defined by

$$V_i(g_i; t) = \frac{C \cdot g_i}{P_i(0)} P_i(t), \tag{1}$$

where $C$ is an initial amount of money for investment at $t = 0$.

In the field of regression analysis, a coefficient of determination or a correlation coefficient has often used as a measure of how well an estimated regression fits. A coefficient of determination is the square of a correlation coefficient. As each coefficient approaches 1, an estimated regression fits better (for this, see, e.g., Downie and Heath [12]). Our index fund consists of the fixed $N$ companies. These companies have already been given by a heuristic rule before the optimization process using a GA and a local search. The heuristic rule (described in Section 3.1) gives $N$ companies that make a similar tendency to the behavior of market index. In this context, the total price of $N$ companies makes a positive linear association to the market index. Hence, we apply not a correlation coefficient but a coefficient of determination to evaluate index funds. The coefficient of determination between the return rates of fund $y(\bar{g}; t)$ and the rates of change of market index $x(t)$ is denoted by

$$R^2(\bar{g}) = \left( \frac{Cov(X, Y)}{\sqrt{Var(X) \cdot Var(Y)}} \right)^2, \tag{2}$$

where $Y = (y(\bar{g}; 1), y(\bar{g}; 2), \cdots, y(\bar{g}; T))$ and $X = (x(1), x(2), \cdots, x(T))$ as an objective. $Cov(X, Y)$ is covariance of $X$ and $Y$. $Var(X)$ and $Var(Y)$ are variance of $X$ and $Y$, respectively.

In this paper, we define the index fund optimization problem as follows:

**Problem:**

$$\max \ R^2(\bar{g})$$
$$\text{s.t.} \ \sum_{i=1}^{N} g_i = 1$$

## 3 Method for the Index Fund Optimization

In this paper, we discuss the Problem defined by Section 2 in order to optimize the proportion of funds in index fund. We propose a method consisting of the following two steps.

**Step 1.** Optimize the proportion of funds in index fund under cost constraints by using a GA that pursues maximizing the coefficient of determination between the fund's return rates and the rates of change of market index.
**Step 2.** Re-optimize the proportion of funds for each company by using a heuristic local search in order that the fund's return rates make a stronger linear association with the rates of change of market index on a scatter diagram.

We introduce each algorithm of the two steps in Sections 3.1 and 3.2, respectively.

### 3.1 Step 1 Based on a GA

Step 1 in our method is to optimize the proportion of funds for index fund under cost constraints by using a GA. This step is based on the methodology of Orito et al. [11].

Suppose that a stock market consists of $K$ listed companies, numbered Companys $1, 2, \cdots, K$. For Company $i$ on the market, the turnover average of company's trading volume $u_i(t)$ multiplied by the price $P_i(t)$ between $t = 1$ and $t = T$ is defined by

$$U_i = \frac{1}{T} \sum_{t=1}^{T} u_i(t) P_i(t) \quad (i = 1, 2, \cdots, K).$$

We renumber the $K$ companies so that

$$U_1 \geq U_2 \geq \cdots \geq U_i \geq \cdots \geq U_K.$$

Note that the renumbered Company $i$ has the $i$-th highest $U_i$ of all companies. Before the optimization process using GA, we choose $N$ companies from renumbered Company 1 to Company $N$. Step 1 in our method applies the following GA to these $N$ companies. For the GA, a gene means the proportion $g_i$ of funds defined by

$$g_i \in [0,1] \quad (i = 1, 2, \cdots, N)$$

and a chromosome means the fund $\bar{g}$ denoted by $\bar{g} = (g_1, g_2, \cdots, g_N)$, where $\sum_{i=1}^{N} g_i$. The fitness value of the GA is the coefficient of determination $R^2(\bar{g})$ given by Equation (2). Our GA is a standard GA. Each operation of the GA is designed as follows.

On the 1st generation of the GA, we generate the initial population at random. In the crossover, we apply the standard two-point crossover for exchanging the partial structure between the two chromosomes and repair to a probability distribution via renormalization. In the mutation, we apply the standard two-point mutation for replacing the partial structure of the selected chromosomes with a new random value in $[0, 1]$ and repair to a probability distribution via renormalization. After making offspring, we apply a roulette wheel selection and an elitism method of one chromosome based on the fitness value. Finally the GA is broken off on the last generation. The population size and generation size are given in the numerical experiments, respectively.

After executing the GA, we select one chromosome with the highest coefficient of determination $R^2(\bar{g})$. The $g_i$ of this chromosome means the proportion of funds for the index fund obtained by Step 1 in our method.

### 3.2 Step 2 Based on a Heuristic Local Search

After applying Step 1, we obtain an index fund with high coefficient of determination. However, it is difficult to find the best solution in large solution spaces by using only a GA as a stochastic searching method. In this context, we apply a heuristic local search as the Step 2 in order to find better index funds than the funds obtained by Step 1.

A sample scatter diagram between the fund's return rates and the rates of change of market index is shown in Figure 1. A black circle represents a data

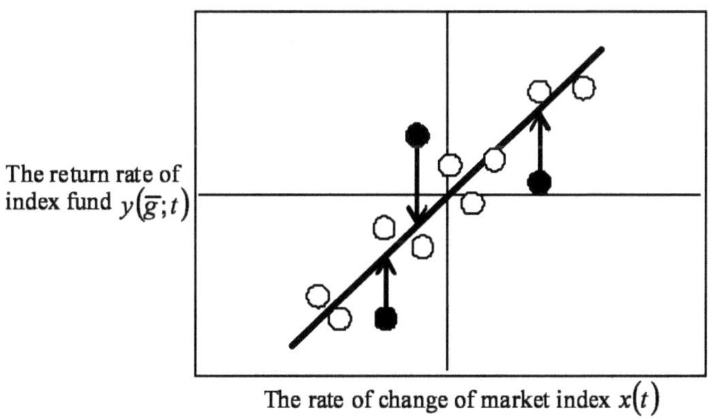


**Fig. 1.** Sample scatter diagram

point at $t$ far from the linear line and a white circle represents a data point at $t$ near from the linear line. It is desired to decrease the distances between all data points and the linear line, but it is difficult because there are a lot of data points on the scatter diagram. Hence, we try to pick up some data points having the large distances as shown by a black circle and try to decrease the distances of these points. We propose the Step 2 that re-optimizes the proportion of funds for companies on the large-distance-data-point in order that the fund's return rates make a stronger linear association with the rates of change of market index on the scatter diagram.

Step 2 in our method applies the following algorithm to the index fund obtained by Step1.

**Step 2-1.** We plot the data point on the rate of change of market index $x(t)$ and the return rate of index fund $y(\bar{g};t)$ at $t$ on the scatter diagram. We then estimate a linear line that summarizes the relationship between the return rates and the rates of change by using regression analysis. The linear line is defined by

$$\hat{y}(\bar{g};t) = \hat{a}x(t) + \hat{b},$$

where $\hat{a}$ and $\hat{b}$ are estimated parameters by the least square method.

When the rate of change $x(t)$ on the horizontal axis is fixed, the distance between $y(\bar{g};t)$ and $\hat{y}(\bar{g};t)$ is defined by

$$E(t) = |y(\bar{g};t) - \hat{y}(\bar{g};t)| \,. \tag{3}$$

Without loss of generality, we renumber the $T$ dates data so that $E(t_1) \geq E(t_2) \geq \cdots \geq E(t_T)$. Note that the renumbered date $t_j$ has the $j$-th largest distance $E(t_j)$ on the scatter diagram.

**Step 2-2.** By Equation (1), we obtain the present value average of $J$ dates for Company $i$,

$$\bar{V}_i = \frac{1}{J}\sum_{j=1}^{J} V_i(g_i;t_j) \qquad (i = 1, 2, \cdots, N).$$

We renumber the $N$ companies so that $V_{i_1} \geq V_{i_2} \geq \cdots \geq V_{i_N}$. Note that the $i_l$ is the $l$-th largest present value average in all companies of $J$ data points with large distances on the scatter diagram.

As the Step 2-3 in our method, we try to re-optimize the renumbered $n$ companies from Company $i_1$ to Company $i_n$ because these $n$ companies seem to have a big impact to make stronger linear association with the rates of change of market index on the scatter diagram.

**Step 2-3.** We re-optimize the proportion of funds for each of $n$ ($\leq N$) companies. The new proportion of funds is defined by

$$g'_{i_l} = \begin{cases} w_{i_l} \cdot g_{i_l} & (l \leq n) \\ g_{i_l} & (\text{otherwise}) \end{cases} \tag{4}$$

$$\text{s.t.} \quad \underset{w=0,0.1,0.2,\cdots,2}{arg\ max} R^2(\bar{g}) \qquad \left(\bar{g} = (g'_{i_1}, \cdots, g'_{i_{l-1}}, wg_{i_l}, g_{i_{l+1}}, \cdots, g_{i_N})\right)$$

where $w_{i_l}$ is a weight parameter for the re-optimization. From Equation (4) with $l = 1$, we can get $g'_{i_1}$, and by using Equation (4) with $l = 2$, we can get $g'_{i_2}$, and in similar manner, $g_{i_3}, \cdots, g_{i_n}$. By this step, we obtain the new index fund whose coefficient of determination is higher than the old one.

The $\bar{g} = (g'_1, \cdots, g'_n, g_{n+1}, \cdots, g_N)$ is the re-optimized index fund obtained by our method consisting of Steps 1 and 2.

## 4 Numerical Experiments

We have conducted numerical experiments to confirm the efficiency of our method. We applied our method to each of 21 data periods on the First Section of Tokyo Stock Exchange consisting of more than 1500 listed companies from Jan. 6, 1997 to Jul. 15, 2005. Each data period contains 100 days, i.e., $t \in [1, 100]$. The data period is shifted every 100 days from Jan. 6, 1997 to Jul. 15, 2005, and it is called from Period 1 (Jan. 6, 1997 – May. 30, 1997) to Period 21 (Feb. 21, 2005 – Jul. 15, 2005). We set the Tokyo Stock Price Index (TOPIX) as a market index. The TOPIX is a composite market index of all common stocks listed on the First Section of Tokyo Stock Exchange and it is basically a measure of the changes in aggregate market value of the market common stocks. The return rates of index fund obtained by our method, therefore, follow a similar path to the rates of change of the TOPIX.

In order to demonstrate the efficiency of our method, we compare the experiments using three methods as follows:

- **GAM1 (GA Method 1)**
  This is a method using a simple GA and it is constructed only by the Step 1 in our method. As mentioned in Section 3.1, maximizing the coefficient of determination is adopted as maximizing the fitness value of GA. The parameters are set as follows:

  The number of genes (companies): $N = 200$,
  The population size: 100,
  The generation size: 100,
  The crossover rate: 0.9,
  The mutation rate: 0.1.
  For fixed $N = 200$ genes, the GA is executed for 20 times.
- **GAM2 (GA Method 2)**
  On the Step 2 in our method, we try to decrease the distances given by Equation (3) in order to make the linear association between the fund's return rates and the rates of change of market index strong. In this method, we use a GA to try to improve this linear association. In this context, we set the total distance $\sum E(t)$ to the objective function for this method.

  This method is constructed only by the GA. Minimizing the total distance is adopted as maximizing the fitness value of GA. The parameters are the same as them of GAM1. For fixed $N = 200$ genes, the GA is executed for 20 times.

**Table 1.** The coefficients of determination obtained by GAM1, GAM2 and GALSM

| Period | Best | | | Worst | | | Average | | | Standard Deviation | | |
|---|---|---|---|---|---|---|---|---|---|---|---|---|
| | GAM1 | GAM2 | GALSM | GAM1 | GAM2 | GALSM | GAM1 | GAM2 | GALSM | GAM1 | GAM2 | GALSM |
| 1 | 0.9710 | 0.9751 | 0.9836 | 0.9592 | 0.9551 | 0.9715 | 0.9649 | 0.9640 | 0.9781 | 1.433E-05 | 3.344E-05 | 1.330E-05 |
| 2 | 0.9729 | 0.9754 | 0.9838 | 0.9521 | 0.9568 | 0.9718 | 0.9625 | 0.9678 | 0.9781 | 2.633E-05 | 1.869E-05 | 8.924E-06 |
| 3 | 0.9866 | 0.9897 | 0.9912 | 0.9813 | 0.9840 | 0.9880 | 0.9836 | 0.9860 | 0.9894 | 1.361E-06 | 1.786E-06 | 4.913E-07 |
| 4 | 0.9776 | 0.9809 | 0.9876 | 0.9691 | 0.9719 | 0.9787 | 0.9736 | 0.9762 | 0.9838 | 8.267E-06 | 5.743E-06 | 6.185E-06 |
| 5 | 0.9906 | 0.9927 | 0.9950 | 0.9866 | 0.9880 | 0.9913 | 0.9885 | 0.9908 | 0.9936 | 1.997E-06 | 2.054E-06 | 9.933E-07 |
| 6 | 0.9685 | 0.9781 | 0.9856 | 0.9577 | 0.9621 | 0.9753 | 0.9634 | 0.9709 | 0.9799 | 1.324E-05 | 2.645E-05 | 8.313E-06 |
| 7 | 0.9484 | 0.9545 | 0.9738 | 0.9058 | 0.9112 | 0.9486 | 0.9315 | 0.9406 | 0.9608 | 1.173E-04 | 1.128E-04 | 5.137E-05 |
| 8 | 0.9436 | 0.9225 | 0.9696 | 0.9002 | 0.8727 | 0.9410 | 0.9217 | 0.8997 | 0.9589 | 1.101E-04 | 2.108E-04 | 4.763E-05 |
| 9 | 0.9687 | 0.9602 | 0.9862 | 0.9432 | 0.9351 | 0.9638 | 0.9521 | 0.9499 | 0.9759 | 5.065E-05 | 4.052E-05 | 4.354E-05 |
| 10 | 0.9751 | 0.9719 | 0.9884 | 0.9482 | 0.9474 | 0.9739 | 0.9605 | 0.9595 | 0.9819 | 5.077E-05 | 4.547E-05 | 2.199E-05 |
| 11 | 0.9853 | 0.9909 | 0.9944 | 0.9736 | 0.9778 | 0.9842 | 0.9796 | 0.9854 | 0.9904 | 8.194E-06 | 1.274E-05 | 5.041E-06 |
| 12 | 0.9897 | 0.9926 | 0.9945 | 0.9825 | 0.9852 | 0.9899 | 0.9863 | 0.9898 | 0.9929 | 5.122E-06 | 3.245E-06 | 1.630E-06 |
| 13 | 0.9792 | 0.9832 | 0.9886 | 0.9708 | 0.9566 | 0.9813 | 0.9754 | 0.9751 | 0.9849 | 5.314E-06 | 4.654E-05 | 4.599E-06 |
| 14 | 0.9926 | 0.9930 | 0.9957 | 0.9829 | 0.9840 | 0.9901 | 0.9863 | 0.9897 | 0.9929 | 5.549E-06 | 4.330E-06 | 2.091E-06 |
| 15 | 0.9860 | 0.9880 | 0.9931 | 0.9794 | 0.9836 | 0.9867 | 0.9830 | 0.9853 | 0.9906 | 5.362E-06 | 1.310E-06 | 3.167E-06 |
| 16 | 0.9819 | 0.9882 | 0.9898 | 0.9640 | 0.9772 | 0.9777 | 0.9736 | 0.9816 | 0.9858 | 2.831E-05 | 8.036E-06 | 1.185E-05 |
| 17 | 0.9800 | 0.9874 | 0.9899 | 0.9706 | 0.9735 | 0.9814 | 0.9754 | 0.9810 | 0.9867 | 8.158E-06 | 1.312E-05 | 5.771E-06 |
| 18 | 0.9814 | 0.9831 | 0.9880 | 0.9614 | 0.9686 | 0.9744 | 0.9700 | 0.9766 | 0.9828 | 1.726E-05 | 1.651E-05 | 1.317E-05 |
| 19 | 0.9901 | 0.9943 | 0.9955 | 0.9872 | 0.9909 | 0.9929 | 0.9888 | 0.9924 | 0.9941 | 5.388E-07 | 1.067E-06 | 5.195E-07 |
| 20 | 0.9798 | 0.9866 | 0.9895 | 0.9685 | 0.9771 | 0.9836 | 0.9753 | 0.9823 | 0.9866 | 7.992E-06 | 1.104E-05 | 3.226E-06 |
| 21 | 0.9817 | 0.9863 | 0.9892 | 0.9750 | 0.9785 | 0.9840 | 0.9781 | 0.9833 | 0.9864 | 3.491E-06 | 4.374E-06 | 2.335E-06 |

**Table 2.** The results of the Wilcoxon rank-sum test

| | GALSM and GAM1 | | | GALSM and GAM2 | | |
|---|---|---|---|---|---|---|
| Period | Statistic $Z$ | P-value | Decision | Statistic $Z$ | P-value | Decision |
| 1 | 5.41 | 6.81E-08 | ** | 5.27 | 1.43E-07 | ** |
| 2 | 5.38 | 7.92E-08 | ** | 5.19 | 2.22E-07 | ** |
| 3 | 5.41 | 6.81E-08 | ** | 4.92 | 9.14E-07 | ** |
| 4 | 5.41 | 6.81E-08 | ** | 5.25 | 1.66E-07 | ** |
| 5 | 5.41 | 6.81E-08 | ** | 4.98 | 6.93E-07 | ** |
| 6 | 5.41 | 6.81E-08 | ** | 4.92 | 9.14E-07 | ** |
| 7 | 5.41 | 6.81E-08 | ** | 5.09 | 3.95E-07 | ** |
| 8 | 5.38 | 7.92E-08 | ** | 5.41 | 6.81E-08 | ** |
| 9 | 5.36 | 9.19E-08 | ** | 5.41 | 6.81E-08 | ** |
| 10 | 5.38 | 7.92E-08 | ** | 5.41 | 6.81E-08 | ** |
| 11 | 5.36 | 9.19E-08 | ** | 4.03 | 5.90E-05 | ** |
| 12 | 5.41 | 6.81E-08 | ** | 4.71 | 2.69E-06 | ** |
| 13 | 5.41 | 6.81E-08 | ** | 5.06 | 4.55E-07 | ** |
| 14 | 5.19 | 2.22E-07 | ** | 4.44 | 9.76E-06 | ** |
| 15 | 5.41 | 6.81E-08 | ** | 5.30 | 1.24E-07 | ** |
| 16 | 5.25 | 1.66E-07 | ** | 3.49 | 5.09E-04 | ** |
| 17 | 5.41 | 6.81E-08 | ** | 4.33 | 1.60E-05 | ** |
| 18 | 5.19 | 2.22E-07 | ** | 3.90 | 1.04E-04 | ** |
| 19 | 5.41 | 6.81E-08 | ** | 4.31 | 1.70E-05 | ** |
| 20 | 5.41 | 6.81E-08 | ** | 3.99 | 7.00E-05 | ** |
| 21 | 5.41 | 6.81E-08 | ** | 4.30 | 1.81E-05 | ** |

**: 99 percent significance.

- **GALSM (GA and Local Search Method)**
  This is our proposing method. We try to maximize the coefficient of determination on the GA process and make the linear association strong on the local search process. The parameters of Step 1 in our method are the same as them of GAM1. The parameters of Step 2 in our method are set as follows:

  The number of dates data for calculating present value average in the Step 2-2: $J = 10$ (about 10% of $T = 100$),
  The number of companies for re-optimization in the Step 2-3: $n = 40$ (20% of $N = 200$).

  We apply the Step 2 to each of 20 funds obtained by the Step 1 using GA.

The best, worst, average and standard deviation of 20 coefficients of determination obtained by GAM1, GAM2 and GALSM are shown in Table 1, respectively.

Table 1 showed that the best coefficients obtained by GALSM were higher than those of GAM1 and GAM2 in all periods. GAM2 gives the higher coefficients than those of GAM1 in Periods 1, 2, 3, 4, 5, 6, 7, 11, 12, 13, 14, 15, 16, 17, 18, 19, 20 and 21. On the other hand, the coefficients of GAM2 are lower than those of GAM1 in Periods 7, 8 and 9. However, using only GAM1 or GAM2 does not give the high coefficients compared with the GALSM.

For each period, the results of GALSM are tested applying a statistical method compared with the results of GAM1 or GAM2. As a statistical method, we use the Wilcoxon rank-sum test which is a non-parametric test for assessing whether two samples of observations come from the same distribution. For this test, one samples are 20 coefficients of determination obtained GALSM and the other samples are 20 coefficients obtained GAM1 or GAM2. The results of the Wilcoxon rank-sum test are shown in Table 2.

From Table 2, the differences between the distributions of the results obtained by GALSM and GAM1 or GAM2 are statistically significant. Hence the numerical experiments confirm the effectiveness of our proposing method, i.e. GALSM.

## 5 Conclusions

In this paper, we have proposed a method for the index fund optimization. This proposed method adopts the GA first and obtains the fund. After using the GA, the method applies the heuristic local search to the fund obtained by the GA. The heuristic local search re-optimizes only the companies having a bad influence on our objective function.

We applied the method to the Tokyo Stock Exchange in the numerical experiments. The results demonstrated the effectiveness of our method compared with other simple GA methods. This means that our local search based on the scatter diagram is effective in optimizing index funds.

In our future works, we will improve our method by proposing a hybrid GA including a heuristic local search, an effective method for the total search and so on.

## References


1. Laws, J., Thompson, J.: Hedging effectiveness of stock index futures. European Journal of Operational Research 163, 177–191 (2005)
2. Elton, E., Gruber, G., Blake, C.: Survivorship bias and mutual fund performance. Review of Financial Studies 9, 1097–1120 (1995)
3. Gruber, M.J.: Another puzzle: the growth in actively managed mutual funds. Journal of Finance 51(3), 783–810 (1996)
4. Malkiel, B.: Returns from investing in equity mutual funds 1971 to 1991. Journal of Finance 50, 549–572 (1995)
5. Xia, Y., Liu, B., Wang, S., Lai, K.K.: A model for portfolio selection with order of expected returns. Computers & Operations Research 27, 409–422 (2000)
6. Chang, T.-J., Meade, N., Beasley, J.E., Sharaiha, Y.M.: A model for portfolio selection with order of expected returns. Computers & Operations Research 27, 1271–1302 (2000)
7. Oh, K.J., Kim, T.Y., Min, S.: Using genetic algorithm to support portfolio optimization for index fund management. Expert Systems with Applications 28, 371–379 (2005)

8. Takabayashi, A.: Selecting and Rebalancing Funds with Genetic Algorithms (in Japanese). In: Proc. of the 1995 Winter Conference of the Japanese Association of Financial Econometrics and Engineering (1995)
9. Aiello, A., Chieffe, N.: International index funds and the investment portfolio. Financial Services Review 8, 27–35 (1999)
10. Chang, K.P.: Evaluating mutual fund performance: an application of minimum convex input requirement set approach. Computers & Operations Research 31, 929–940 (2004)
11. Orito, Y., Takeda, M., Iimura, K., Yamazaki, G.: Evaluating the Efficiency of Index Fund Selections Over the Fund's Future Period. Computational Intelligence in Economics and Finance 2, 157–168 (2007)
12. Downie, N.M., Health, R.W.: Basic Statistical Methods. Harper and Row, New York (1983)

# Mining Bayesian Networks from Direct Marketing Databases with Missing Values

Yuan Yuan Guo[1] and Man Leung Wong[2]

[1] Department of Computing and Decision Sciences, Lingnan University, Tuen Mun, Hong Kong
bluegirlgyy@gmail.com

[2] Department of Computing and Decision Sciences, Lingnan University, Tuen Mun, Hong Kong
mlwong@ln.edu.hk

Discovering knowledge from huge databases with missing values is a challenging problem in Data Mining. In this paper, a novel hybrid algorithm for learning knowledge represented in Bayesian Networks is discussed. The new algorithm combines an evolutionary algorithm with the Expectation-Maximization (EM) algorithm to overcome the problem of getting stuck in sub-optimal solutions which occurs in most existing learning algorithms. The experimental results on the databases generated from several benchmark network structures illustrate that our system outperforms some state-of-the-art algorithms. We also apply our system to a direct marketing problem, and compare the performance of the discovered Bayesian networks with the response models obtained by other algorithms. In the comparison, the Bayesian networks learned by our system outperform others.

## 1 Introduction

In real-life applications, the collected databases may contain missing values in the records. Irrelevant records or trivial items with missing values can be simply discarded from the raw databases in the data preprocessing procedure. However, in most cases, the variables are related to each other and the deletion of incomplete records may lose important information. This will affect performance dramatically especially if we want to discover some knowledge "nuggets" from the databases and they happen to be contained in the incomplete records. Usually, people may alternatively replace the missing values with certain values, such as the mean or mode of the observed values of the same variable. Nevertheless, it may change the distribution of the original database.

Bayesian networks are popular within the community of artificial intelligence and data mining due to their ability to support probabilistic reasoning from data with uncertainty. They can represent the relationships among random variables and the conditional probabilities of each variable from a given database. With a network structure at hand, people can conduct probabilistic inference to predict the outcome of some variables based on the values of other observed ones. Hence,

M. Gen et al.: Intelligent and Evolutionary Systems, SCI 187, pp. 13–35.
springerlink.com 

Bayesian networks are widely used in many areas, such as diagnostic and classification systems [1, 2, 3], information retrieval [4], troubleshooting [5], and so on. They are also suitable for knowledge reasoning with incomplete information.

Currently, people focus on two kinds of Bayesian network learning problems. In parameter learning, the values of parameters of a known Bayesian network structure are estimated. On the other hand, Bayesian network structures are discovered from databases in structure learning. Many methods have been suggested to learn Bayesian network structures from complete databases without missing values, which can be classified into two main categories [6]: the dependency analysis [7] and the score-and-search approaches [8, 9, 10]. For the former approach, the results of dependency tests are employed to construct a Bayesian network conforming to the findings. For the latter one, a scoring metric is adopted to evaluate candidate network structures while a search strategy is used to find a network structure with the best score. Decomposable scoring metrics, such as Minimum Description Length (MDL) and Bayesian Information Criterion (BIC), are usually used to deal with the time-consuming score evaluation problem. When the network structure changes, we only need to re-evaluate the scores of the corresponding nodes related to the changed edges, rather than the scores of all nodes. Stochastic search methods such as Genetic Algorithms (GAs) [11, 12], Evolutionary Programming (EP) [13], and Hybrid Evolutionary Algorithm (HEA) [14] have also been proposed in the score-and-search approach. They demonstrated good performance in learning Bayesian networks from complete databases.

Nevertheless, learning Bayesian networks from incomplete databases is a difficult problem in real-world applications. The parameter values and the scores of networks cannot be computed directly on the records having missing values. Moreover, the scoring metric cannot be decomposed directly. Thus, a local change in the network structure will lead to the re-evaluation of the score of the whole network structure, which is time-consuming considering the number of all possible network structures and the complexity of the network structures. Furthermore, the patterns of the missing values also affect the dealing methods. Missing values can appear in different situations: *Missing At Random*, or *Not Ignorable* [15]. In the first situation, whether an observation is missing or not is independent of the actual states of the variables. So the incomplete databases may be representative samples of the complete databases. However, in the second situation, the observations are missing to some specific states for some variables. Different approaches should be adopted for different situations, which again complicates the problem.

Many researchers have been working on parameter learning and structure learning from incomplete databases. For the former, several algorithms, such as Gibbs sampling, EM [9], and *Bound-and-Collapse* (BC) method [16, 17], can be used to estimate or optimize the parameter values for a Bayesian network whose structure has been given or known. For structure learning from incomplete databases, the main issues are how to define a suitable scoring metric and how to search for Bayesian network structures efficiently and effectively. Concerning the score evaluation for structure learning, some researchers proposed calculating the expected values of the statistics to approximate the score of candidate networks. Friedman proposed a Bayesian *Structural Expectation-Maximization* (SEM) algorithm which alternates between the parameter optimization process and the model search process [18, 19]. The score of a Bayesian network is maximized by means of the maximization of the expected score. Peña *et al.* used the BC+EM method instead of the EM method in their Bayesian Structural BC+EM algorithm for clustering [20, 21]. However, the search strategies adopted in most existing SEM algorithms may not be effective and may make the algorithms find sub-optimal solutions. Myers *et al.* employed a genetic algorithm to learn Bayesian networks from incomplete databases [22]. Both network structures and the missing values are encoded and evolved. The incomplete databases are completed by specific genetic operators during evolution. Nevertheless, it has the efficiency and convergence problems because of the enlarged search space and the strong randomness of the genetic operators for completing the missing values.

In this paper, we propose a new data mining system that uses EM to handle incomplete databases with missing values and uses a hybrid evolutionary algorithm to search for good candidate Bayesian networks. The two procedures are iterated so that we can continue finding a better model while optimizing the parameters for a good model to complete the database with more accurate information. Instead of using the expected values of statistics as in most existing SEM algorithms, our system applies a data completing procedure to complete the database and thus decomposable scoring metrics can be used to evaluate the networks. The MDL scoring metric is employed in the search process to evaluate the fitness of the candidate networks. In this study, we consider the situations that the unobserved data are missing at random. We also demonstrate that our system outperforms some state-of-the-art algorithms.

The rest of this paper is organized as follows. In Section 2, we will present the backgrounds of Bayesian networks, the missing value problem, and some Bayesian network learning algorithms. In Section 3, our new data mining system for incomplete databases, *HEAm*, will be described in details. A number of experiments have been conducted to compare our system with other learning algorithms and the results will be discussed in Section 4. In Section 5, we use our system to discover Bayesian networks from a real-life direct marketing database. We will conclude the paper in the last section.

## 2 Background

### 2.1 Bayesian Networks

A Bayesian network has a directed acyclic graph (DAG) structure. Each node in the graph represents a discrete random variable in the domain. An edge, $Y \rightarrow X$, on the graph, describes a parent-child relation in which $Y$ is the parent and $X$ is the child. All parents of $X$ constitute the parent set of $X$ which is denoted by $\Pi_X$. In addition to the graph, each node has a conditional probability table specifying the probability of each possible state of the node given each possible combination of states of its parents. For a node having no parent, the table gives the marginal probabilities of the node.

Let $U$ be the set of variables in the domain, i.e., $U = \{X_1,\ldots,X_n\}$. Following Pearl's notation [23], a conditional independence (CI) relation is denoted by $I(X, Z, Y)$ where $X$, $Y$, and $Z$ are disjoint subsets of variables in $U$. Such notation says that $X$ and $Y$ are conditionally independent given the *conditioning set* $Z$. Formally, a CI relation is defined with:

$$P(x \mid y, z) = P(x \mid z) \quad \text{whenever} \quad P(y, z) > 0 \tag{1}$$

where $x$, $y$, and $z$ are any value assignments to the set of variables $X$, $Y$, and $Z$ respectively. For a Bayesian network, the CI relation can be understood as: given the states of its parents, each node is conditionally independent of its non-descendants in the graph. A CI relation is characterized by its *order*, which is the number of variables in $Z$. When $Z$ is $\emptyset$, the order is 0.

By definition, the joint probability distribution of $U$ can be expressed as:

$$P(X_1, \ldots, X_n) = \prod_i P(X_i | \Pi_{X_i}) \tag{2}$$

For simplicity, we use $X_i = k$ to specify that the $i$-th node takes the $k$-th possible state in its value domain, $\Pi_{X_i} = j$ to represent $\Pi_{X_i}$ being instantiated to the $j$-th combinational state, and $N_{ijk}$ to represent the counts of $X_i = k$ and $\Pi_{X_i} = j$ appearing simultaneously in the database. The conditional probability $p(X_i = k|\Pi_{X_i} = j)$, also denoted as *parameter* $\theta_{ijk}$, can be calculated from the complete database by $\theta_{ijk} = \frac{N_{ijk}}{\sum_k N_{ijk}}$.

### 2.2 Bayesian Network Structure Learning

As mentioned above, there are two main categories of Bayesian network structure learning algorithms. The dependency analysis approach constructs a network by testing the validity of any independence assertion $I(X, Z, Y)$. If the assertion is supported by the database, edges cannot exist between $X$ and $Y$ on the graph. The validity of $I(X, Z, Y)$ is tested by performing a CI-test, and statistical hypothesis testing procedure could be used. Suppose that the likelihood-ratio $\chi^2$ test is used and the $\chi^2$ statistics is calculated by

$$\begin{aligned} g^2 &= -2 \sum_{x,y,z} P(x, y, z) * \log \frac{P(x, y, z)}{P(y, z)P(x|y, z)} \\ &= -2 \sum_{x,y,z} P(x, y, z) * \log \frac{P(x, y, z)}{P(y, z)P(x|z)} \end{aligned} \tag{3}$$

Checking the computed $g^2$ against the $\chi^2$ distribution, we can obtain the $p$-value [14]. If the $p$-value is less than a predefined *cutoff value* $\alpha$, the assertion $I(X, Z, Y)$ is not valid; otherwise, it is valid and edges cannot exist between $X$ and $Y$. Hence, network structures can be constructed according to the test results.

The score-and-search approach adopts some scoring metrics to evaluate the candidate networks and uses some search strategies to find better network structures. Due to the large number of possible network structures and the complexity of structures, it is time-consuming to re-evaluate each network structure once its structure changes. Hence, decomposable scoring metrics are commonly used to tackle the score evaluation problem. Take MDL scoring metric for example, the MDL score of network $G$ with every node $N_i$ in the domain $U$ can be described as $MDL(G)=\sum_{N_i \in U} MDL(N_i, \Pi_{N_i})$. Since the metric is node-decomposable, it is only necessary to re-calculate the MDL scores of the nodes whose parent sets have been modified when the network structure changes, while scores of other nodes can be re-used in the searching procedure. With a scoring metric, the learning problem becomes a search problem. Various search strategies have been applied for the problem.

### 2.3 HEA

HEA is a score-and-search method, which is proposed for learning Bayesian networks from complete databases [14]. It employs the results of lower order (order-0 and order-1) CI-tests to refine the search space and adopts a hybrid evolutionary algorithm to search for good network structures. Each individual in the population represents a candidate network structure which is encoded by a connection matrix. Besides, each individual has a cutoff value $\alpha$ which is also subject to be evolved. At the beginning, for every pair of nodes (X,Y), the highest $p$-value returned by the lower order CI-tests is stored in a matrix $P_v$. If the $p$-value is greater than or equal to $\alpha$, the conditional independence assertion $I(X,Z,Y)$ is assumed to be valid, which implies that nodes $X$ and $Y$ cannot have a direct edge between them. By changing the values of $\alpha$ dynamically, the search space of each individual can be modified and each individual conducts its search in a different search space. Four mutation operators are specifically designed in HEA. They add, delete, move, or reverse edges in the network structures either through a stochastic method or based on some knowledge. A novel merge operator is suggested to reuse previous search results. The MDL scoring metric is used for evaluating candidate networks. A cycle prevention method is adopted to prevent cycle formation in the network structures. The experimental results in [14] demonstrated that HEA has better performance on some benchmark databases and real-world databases than other state-of-the-art algorithms.

### 2.4 The Missing Value Problem

In real-world applications, the databases may contain incomplete records which have missing values. People may simply discard incomplete records, but relevant information may be deleted. Alternatively, they can complete the missing values with the information of the databases such as the mean values of other observed values of the variables. However, the distribution of the data may be changed.

One advantage of Bayesian networks is that they support probabilistic reasoning from data with uncertainty. However, for learning Bayesian networks from incomplete databases, the parameter values and the scores of networks cannot be computed directly on the incomplete records which contain missing values. Besides, the decomposable scoring metric cannot be applied directly. Thus, a local change in the network structure will lead to the re-evaluation of the score of the whole network structure.

For parameter learning, existing methods either complete the missing values or use different inference algorithms to get the expected values of statistics. Two commonly adopted methods are Gibbs sampling and EM [9]. Gibbs sampling tries to complete the database by inferring from the available information and then learns from the completed database. On the other hand, EM calculates the expected values of the statistics via inference and then updates the parameter values using the previously calculated expected values [24, 25]. It will converge to a local maximum of the parameter values under certain conditions. Furthermore, EM usually converges faster than Gibbs sampling. Both Gibbs sampling and EM assume that the missing values appear randomly or follow a certain distribution. In order to encode prior knowledge of the pattern of missing data, Ramoni *et. al.* proposed a new deterministic *Bound-and-Collapse* (BC) method that does not need to guess the pattern of missing data [16, 17]. It firstly bounds the possible estimate consistent with the probability interval by computing the maximum and minimum estimates that would have been inferred from all possible completions of the database. Then the interval is collapsed to a unique value via a convex combination of the extreme estimates using information on the assumed pattern of missing data.

For structure learning from incomplete databases, the score-and-search approach can still be employed. The main issues are how to define a suitable scoring metric and how to search for Bayesian networks efficiently and effectively. Many variants of Structural EM (SEM) algorithm were proposed for this kind of learning in the past few years [18, 19, 20].

The basic SEM algorithm was proposed by Friedman for learning Bayesian networks in the presence of missing values and hidden variables [18]. It alternated between two procedures: an optimization for the Bayesian network parameters conducted by the EM algorithm, and a search for a better Bayesian network structure using a greedy hill climbing strategy. The two procedures iterated until the whole algorithm is stopped. The score of a Bayesian network was approximated by the expected value of statistics. Friedman extended his SEM to directly optimize the true Bayesian score of a network in [19]. The framework of the basic SEM algorithm can be described as follows:

1. let $M_1$ be the initial model structure.
2. for t=1,2,...
   - Execute EM to approximate the maximum-likelihood parameters $\Theta_t$ for $M_t$.
   - Perform a greedy hill-climbing search over model structures, evaluating each model using approximated score $Score(M)$.
   - let $M_{t+1}$ be the model structure with the best score.
   - If $Score(M_t)$ =$Score(M_{t+1})$ then return $M_t$ and $\Theta_t$.

# 3 HEAm for Incomplete Databases

## 3.1 HEAm

Although HEA outperforms some existing approaches, it cannot deal with incomplete databases. A novel data mining system called *HEAm* is developed, which applies EM to deal with missing values in the database and employs HEA to search for good Bayesian networks effectively and efficiently. HEAm is described in Fig. 1.

There are two special kinds of generations in HEAm. *SEM generation* refers to one iteration in the SEM framework (step 12 of Fig. 1) while *HEA iteration* refers to the iteration in HEA search process (step 12(g) of Fig. 1).

In the data preprocess phase, the database is separated and stored into two parts. The set of records having missing values is marked as $H$, and the set of records without missing values is marked as $O$. Order-0 and order-1 CI tests are then conducted on $O$ and the results are stored for refining the search space of each individual in the following procedures.

At the beginning of the SEM phase, for each individual, we check a randomly generated $\alpha$ value with the stored CI-test results to refine the search space of this individual. A DAG structure is then randomly constructed from the refined search space for this individual. Thus, the initial population is generated. The current best network with the best score, denoted as $G_{best}$, is selected from the population after the initial network structures are evaluated on $O$. HEAm will then be executed for a number of SEM generations until the stopping criteria are satisfied. Within each *SEM generation*, EM will be conducted first to find the best parameter values of $G_{best}$ (step 12(a) of Fig. 1). The missing values in $H$ will be filled according to $G_{best}$ and its parameters (step 12(c) of Fig. 1). Combining the newly completed result of $H$ with $O$, we get a new complete data $O'$. Then, HEA search procedure will be executed on $O'$ for a certain number of HEA generations to find a better network structure to replace $G_{best}$. The MDL scoring metric is again employed in the search process to evaluate the fitness of the candidate networks. The whole process will iterate until the maximum number of SEM generations is reached or the log-likelihood of $G_{best}$ doesn't change for a specified number of SEM generations. The log-likelihood of $G_{best}$ in the $t$-th SEM generation can be computed by $ll(G_{best}(t)) = \sum_{i,j,k} [E(N_{ijk}) log(\theta_{ijk})]$. Finally, the best network will be returned. Some techniques are depicted in following subsections.

## 3.2 The EM Procedure in HEAm

EM is employed here for parameter estimation of the current best network $G_{best}$ which will be used for the data completing procedure. In order to facilitate the converge of the EM procedure shown in Fig. 2, we choose the current best network $G_{best}$ as the input network structure. The initial parameter values of $G_{best}$ are computed on data $O*$. For the first execution of EM in the first SEM generation, $O$ is used as $O*$. In the other SEM generations, $O*$ is the completed data $O'$ from the previous SEM generation.

---

**Data Preprocess**

1. Store incomplete records together, mark the whole set as $H$.
2. Store other records together, mark the whole set as $O$.

**CI test Phase**

3. Perform order-0 and order-1 CI tests on $O$.
4. Store the highest $p$-value in the matrix $P_v$.

**SEM phase**

5. Set $t$, the generation count, to 0.
6. Set $t_{SEM}$, the SEM generation count, to 0.
7. Set $t_{uc}$, the count of generations with unchanged log-likelihood, to 0.
8. Initialize the value of $m$, the population size.
9. For each individual $G_i$ in the population $Pop(t)$
   - Initialize the $\alpha$ value randomly.
   - Refine the search space by checking the $\alpha$ value against the $P_v$ value.
   - Inside the reduced search space, create a DAG randomly.
10. Each DAG in the population is evaluated using the MDL metric on current complete data $O$.
11. Pick up the network with the lowest MDL score from $Pop(t)$ as $G_{best}$.
12. While $t_{SEM}$ is less than the maximum number of SEM generations or $t_{uc}$ is less than $MAX_{uc}$,
    a) Execute **EM procedure**.
    b) If the log-likelihood of $G_{best}$ doesn't change, increment $t_{uc}$ by 1;
       else set $t_{uc}$ to 0.
    c) Complete missing data in $H$ using $G_{best}$ and its parameters, and get updated complete data $O'$.
    d) Execute order-0 and order-1 CI-tests on $O'$, and store the highest $p$-value in $P_v$.
    e) For each individual $G_i$ in the population $Pop(t)$
       - Refine the search space by checking the $\alpha$ value against the $P_v$ value.
       - Evaluate $G_i$ using the MDL metric on $O'$.
    f) Set $t_{HEA}$, the HEA generation count in each SEM generation, to 0.
    g) While $t_{HEA}$ is less than the maximum number of HEA generations in each SEM generation ,
       - execute **HEA search procedure**.
       - increase $t_{HEA}$ and $t$ by 1, respectively.
    h) Pick up the individual that has the lowest MDL score on $O'$ to replace $G_{best}$.
    i) Increase $t_{SEM}$ and $t$ by 1, respectively.
13. Return the individual that has the lowest MDL score in any HEA generation of the last SEM generation as the output of the algorithm.

---

**Fig. 1.** The algorithm of HEAm

The EM procedure contains two steps: the E-step and the M-step. In the E-step, the expected values of statistics of unobserved data (often called sufficient statistics) are estimated using probabilistic inference based on the input $G_{best}$

---

**Procedure EM**($G_{best}$, $O*$, $H$)

1. Calculate the parameter values of $G_{best}$ on data $O*$.
2. Set $t$, the EM iteration count, to 0.
3. While not converged,
   - E-step: calculate the expected statistics on $H$ for every node $N_i$.
   - M-step: update $\theta_{ijk}$ using $E'(N_{ijk})$.
   - Calculate the log-likelihood of $G_{best}$.
   - Increase $t$ by 1.
4. Output $G_{best}$ and its parameters.

---

**Fig. 2.** Pseudo-code of the EM procedure

and its parameter assignments. For each node $X_i$ and record $l^*$, we can calculate the expected value of $N_{ijk}$ using the following equation:

$$E(N_{ijk}) = \sum_{l*}^{H} E(N_{ijk}|l^*) \tag{4}$$

where $E(N_{ijk}|l^*) = p(X_i = k, \Pi_{X_i} = j|l^*)$. Let $l$ represents the set of all other observed nodes in $l^*$. When both $X_i$ and $\Pi_{X_i}$ are observed in $l^*$, the expected value can be counted directly which is either 0 or 1. Otherwise, $p(X_i = k, \Pi_{X_i} = j|l^*) = p(X_i = k, \Pi_{X_i} = j|l)$, and it can be calculated using any Bayesian inference algorithm. In our experiments, the junction tree algorithm is employed for Bayesian inference [26]. Since the database is preprocessed, we just need to run the E-step on $H$.

Then, in the M-step, the parameters $\theta_{ijk}$ are updated by

$$\theta_{ijk} = \frac{E'(N_{ijk})}{\sum_k E'(N_{ijk})} \tag{5}$$

where $E'(N_{ijk})$ is the sum of the sufficient statistics calculated on $H$ in the E-step and the statistics calculated on $O$ which are evaluated and stored at the beginning.

The two steps will iterate until either the value of the log-likelihood doesn't change in two successive iterations, or the maximum number of iterations is reached.

### 3.3 Data Completing Procedure

One of the main problems in learning Bayesian networks from incomplete databases is that the node-decomposable scoring metric cannot be used directly. In order to utilize HEA in our data mining system, we complete the missing data after each execution of the EM procedure so that the candidate networks can be evaluated efficiently on a complete database.

When more than one node are unobserved in a record, we fill the missing data according to the topological order of the current best network $G_{best}$. For example, if node $X_i$ and $X_j$ are both unobserved in record $l^*$ and $X_i \rightarrow X_j$ exists in $G_{best}$, we first fill the value of $X_i$ and put it back into the junction tree, and then find a value for $X_j$.

For each missing value, Bayesian inference algorithms are again employed to obtain the probability of each possible state of the unobserved node under the current observed data. Suppose the value of node $X_i$ is unobserved in current record $l^*$, and $X_i$ has $k$ possible states in its value domain. We use $\{p_1, p_2,...,p_k\}$ to represent the set of the inferred probability of each of its state appearing under current observed data in $l^*$, respectively. We can simply pick up the state having the highest probability to replace the missing value. Alternatively, we can select a state via a roulette wheel selection method. In the latter approach, a random decimal $r$ between 0 and 1 is generated, and then the $m$-th state will be chosen if $m = 1$ and $r \leq p_1$, or, $1 < m \leq k$ and $\sum_{i=1}^{m-1} p_i < r \leq \sum_{i=1}^{m} p_i$. In HEAm, we adopt the second completing approach so that the states with lower probabilities may also be selected.

As mentioned in Section 1, one common method for completing the incomplete databases is to replace the missing values with certain values, such as the mean or mode of the observed values of the same variable. The difference between this method and our data completing method is that, we consider the correlations of distributions among all the variables, while the other one only considers the distribution of a single variable. Hence, when discovering knowledge from the whole database, using our method may lead to better models to represent the interrelations among the variables.

### 3.4 HEA Search Procedure

With a complete data set $O'$, the decomposable scoring metrics can be applied and HEA can be utilized to learn good Bayesian networks. The lower order CI-test will be conducted again on $O'$ and the highest $p$-values are stored in the matrix $P_v$, just as mentioned in subsection 2.3. Hence, each individual will refine its search space according to the new information from the new data set $O'$. The candidate networks are evaluated on $O'$ using the MDL scoring metric. In each *HEA iteration*, the mutation operators and the merge operator will be applied on each individual to generate a new offspring. The old individuals and their offspring are selected by the tournament selection procedure to form the new population for the next HEA iteration. The values of $\alpha$ evolve in each iteration, and thus each individual conducts its search in a dynamic search space. This HEA search process continues until the maximum number of HEA iterations have been performed. Finally, the best network will be returned.

## 4 Experiments

### 4.1 Methodology

We compare the performance of HEAm with LibB [27] and Bayesware Discoverer [28] on 12 databases with different sizes and missing percentages. Firstly, we randomly sample three original databases from the well-known benchmark networks including the ALARM, the PRINTD, and the ASIA networks, with no missing values. Then, the 12 incomplete databases used in our experiments are generated from the corresponding original databases with different percentage of missing values introduced randomly. Table 1 depicts the original networks used to generate the original databases, the sizes of the databases, the numbers of nodes, the MDL scores evaluated on the original databases with the original networks, and the source of the databases. Table 2 summarizes the percentage of missing values, the number of missing values (which is equal to *size * nodes * missing percentage*), and the number of incomplete records contain missing values for each incomplete database.

LibB is developed by Friedman and Elidan to learn Bayesian networks from databases in the presence of missing values and hidden variables [27]. By default, LibB applies the Bayesian Dirichlet score metric (BDe), the junction tree inference algorithm, and the greedy hill climbing search method. Some other search methods are also implemented in LibB. For each database, different configurations of the parameter settings are tried, and then LibB is executed with the most appropriate configuration on the database.

Bayesware Discoverer (BayD, for simplicity in this paper) is a software for the Windows environment that builds Bayesian networks from databases [28]. It

**Table 1.** The original databases

| Database | original network | size | nodes | MDL score | source |
|---|---|---|---|---|---|
| Asia10000_o | ASIA | 10000 | 8 | 32531.9 | Netica [29] |
| Printd5000 | PRINTD | 5000 | 26 | 106541.6 | HEA [14] |
| Alarm_o | ALARM | 10000 | 37 | 138455 | HEA [14] |

**Table 2.** The incomplete databases

| Database | missing percent(%) | no. of missing values | no. of incomplete records |
|---|---|---|---|
| Asia10000_o_p0.1 | 0.1 | 80 | 79 |
| Asia10000_o_p1 | 1 | 800 | 777 |
| Asia10000_o_p5 | 5 | 4000 | 3337 |
| Asia10000_o_p10 | 10 | 8000 | 5704 |
| Printd5000_o_p0.1 | 0.1 | 130 | 129 |
| Printd5000_o_p1 | 1 | 1300 | 1133 |
| Printd5000_o_p5 | 5 | 6500 | 3708 |
| Printd5000_o_p10 | 10 | 13000 | 4660 |
| Alarm_o_p0.1 | 0.1 | 370 | 366 |
| Alarm_o_p1 | 1 | 3700 | 3134 |
| Alarm_o_p5 | 5 | 18500 | 8484 |
| Alarm_o_p10 | 10 | 37000 | 9788 |

applies a deterministic *Bound-and-Collapse* method to determine the pattern of missing data [16, 17].

For HEAm, the maximum number of iterations in EM is 10, the maximum number of HEA iterations in each SEM generation is 100, the maximum number of SEM generations is 50, the population size is 50, tournament size is 7, and $MAX_{uc}$ is set to 10.

Since HEAm and LibB are stochastic, we execute them for 40 times on each database to get their average performance. BayD is executed once on each databases because it is deterministic. All of the experiments are conducted on the same PC with a Pentium$^{(R)}$ IV 2.6GHz processor and 512 MB memory running Windows XP operating system.

The performance of the algorithms are evaluated using the following four measures:

1. ASD: the average structural difference, i.e., number of edges added, reserved and omitted, between the final solution and the original network.
2. AESD: the average structural difference between the equivalence class of the final solution with that of the original network. Two Bayesian networks are equivalent if and only if they have the same skeletons and the same $v$-structures [30]. Bayesian networks in the same equivalence class will have the same MDL score on the same database.
3. AET: the average execution time of each trial in seconds.
4. AOMDL: the average MDL score of the final solutions evaluated on the original database. The smaller the score, the better the network structure.

### 4.2 Performance Comparison among Different Methods

In Table 3, the performance comparisons among different algorithms on each incomplete database are summarized. Since HEAm and LibB are executed for 40 times for each database, the figures are the average and the standard deviations of 40 trials. It can be observed that HEAm can always find better network structures with smaller structural difference than BayD. We can also see that HEAm and LibB find the same network structures for the PRINTD databases. From ASD, AESD, and AOMDL of HEAm and LibB for the ASIA and the ALARM databases, it can be observed that HEAm finds better network structures for the two benchmark networks with different percentages of missing values. The differences are significant at 0.05 level using the Mann-Whitney test [31].

To compare the best final solutions found, Table 4 summarizes the best final network structures obtained among the 40 trials according to structural difference (SD) and equivalent structure difference (ESD). Numbers in parentheses are the frequencies of the same network with the best SD or ESD appearing in the total 40 trials. The results on the PRINTD databases are equal and thus omitted in the table. It can be seen that HEAm can find the networks with ESD of 1 in all trials for the three ASIA databases. For the ALARM databases, HEAm has a higher chance of obtaining the recorded best final network, except for the alarm_o_p1 database. For this database, HEAm can find the best network with ESD equal to 2 for 20 times out of the 40 trials, while LibB can only find its best network once. Moreover, HEAm gets a much better average results. We can conclude that HEAm is more stable and effective than LibB on the experimental databases.

**Table 3.** The Performance Comparison Among HEAm, LibB, and BayD

| Database | | ASD | AESD | AET | AOMDL |
|---|---|---|---|---|---|
| Asia10000_o_p0.1 | HEAm | 1.3±0.5 | 1± 0.0 | 2.1±0.3 | 32509.9±0.0 |
| | LibB | 3±0.0 | 2±0.0 | 1±0.0 | 32557.7±0.0 |
| | BayD | 4 | 4 | 9 | 32579.1 |
| Asia10000_o_p1 | HEAm | 1.6±0.5 | 1±0.0 | 3.7±0.7 | 32509.9±0.0 |
| | LibB | 3.7±1.3 | 2.7±1.4 | 1.2±0.4 | 32565.7±17.7 |
| | BayD | 16 | 19 | 15 | 34251.7 |
| Asia10000_o_p5 | HEAm | 1.6±0.5 | 1±0.0 | 11.9±1.6 | 32509.9±0.0 |
| | LibB | 3±0.0 | 2±0.0 | 2.9±0.3 | 32557.7±0.0 |
| | BayD | 18 | 22 | 14 | 35199.1 |
| Asia10000_o_p10 | HEAm | 1.7±0.5 | 1±0.0 | 21.3±0.7 | 32509.9±0.0 |
| | LibB | 3±0.0 | 2±0.0 | 6±0.0 | 32557.7±0.0 |
| | BayD | 18 | 22 | 11 | 35199.1 |
| Printd5000_p0.1 | HEAm | 0±0.0 | 0±0.0 | 75.3±1.2 | 106542±0.0 |
| | LibB | 0±0.0 | 0±0.0 | 51±0.0 | 106542±0.0 |
| | BayD | 49 | 51 | 78 | 106873 |
| Printd5000_p1 | HEAm | 0±0.0 | 0±0.0 | 91.6±2.2 | 106542±0.0 |
| | LibB | 0±0.0 | 0±0.0 | 80±0.0 | 106542±0.0 |
| | BayD | 76 | 78 | 128 | 108141 |
| Printd5000_p5 | HEAm | 0±0.0 | 0±0.0 | 157.3±4.6 | 106542±0.0 |
| | LibB | 0±0.0 | 0 ±0.0 | 189.0±5.1 | 106542±0.0 |
| | BayD | 110 | 114 | 165 | 112898 |
| Printd5000_p10 | HEAm | 0±0.0 | 0±0.0 | 214.8±7.6 | 106542±0.0 |
| | LibB | 0±0.0 | 0 ±0.0 | 319.2±28.6 | 106542±0.0 |
| | BayD | 106 | 112 | 120 | 118860 |
| Alarm_o_p0.1 | HEAm | 8.2±5.7 | 7.8±7.3 | 336.3±64.4 | 138670.1±591.4 |
| | LibB | 31.0±14.8 | 30.9±16.3 | 392.3±93.8 | 142536.3±3642.5 |
| | BayD | 135 | 136 | 595 | 173279 |
| Alarm_o_p1 | HEAm | 5.8±3.4 | 4.7±4.4 | 837.1±460.2 | 138397.1±198.8 |
| | LibB | 29.5±15.9 | 29.7±17.0 | 823.5±232.9 | 142635.9±3372.1 |
| | BayD | 144 | 144 | 650 | 185360 |
| Alarm_o_p5 | HEAm | 6.3±4.1 | 5.7±5.7 | 3291.7±2654.3 | 138399.5±179.3 |
| | LibB | 32.0±10.1 | 32.2±10.7 | 3038.2±799.3 | 142486.2±2251.0 |
| | BayD | 201 | 201 | 1035 | 370700 |
| Alarm_o_p10 | HEAm | 6.8±4.1 | 5.9±5.3 | 6957.3±10566.1 | 138506.8±260.3 |
| | LibB | 47.3±7.3 | 50±8.3 | 4359.5±1300.7 | 143013.3±286.4 |
| | BayD | 62 | 63 | 1920 | 6.31286e+006 |

**Table 4.** The Best final solutions in the 40 trials

| Database | HEAm | | | LibB | | |
|---|---|---|---|---|---|---|
| | SD | ESD | OMDL | SD | ESD | OMDL |
| Asia10000_o_p0.1 | 1(25) | 1(40) | 32509.9 | 3(40) | 2(40) | 32557.7 |
| Asia10000_o_p1 | 1(18) | 1(40) | 32509.9 | 3(31) | 2(31) | 32557.7 |
| Asia10000_o_p5 | 1(17) | 1(40) | 32509.9 | 3(40) | 2(40) | 32557.7 |
| Asia10000_o_p10 | 1(14) | 1(40) | 32509.9 | 3(40) | 2(40) | 32557.7 |
| Alarm_o_p0.1 | 3(6) | 2(21) | 138275 | 5(1) | 2(1) | 138481 |
| Alarm_o_p1 | 2(2) | 2(20) | 138275 | 1(1) | 0(1) | 138455 |
| Alarm_o_p5 | 2(1) | 2(25) | 138275 | 12(1) | 11(1) | 138982 |
| Alarm_o_p10 | 2(4) | 2(24) | 138275 | 12(1) | 13(1) | 140701 |

# 5 Application in a Real-World Problem

In this section, HEAm is applied to a real-world data mining problem. The problem relates with direct marketing in which the objective is to predict and rank potential buyers from the buying records of previous customers. The customer list will be ranked according to each customer's likelihood of purchase [32, 33]. The decision makers can then select the portion of customer list to roll out. An advertising campaign including mailing of catalogs or brochure is targeted on the most promising prospects. Hence, if the prediction is accurate, it can help to enhance the response rate of the advertising campaign and increase the return of investment. Since Bayesian networks can estimate the posterior probability of an instance (a customer) belonging to a particular class (active or non-active respondents), by assuming that the estimated probability is equal to the likelihood of buying, Bayesian networks are particularly suitable to be the response model. Therefore, we will learn Bayesian networks from the real-world databases as response models, and evaluate the performance of different models from a perspective of direct marketing.

## 5.1 The Direct Marketing Problem

Direct marketing concerns communication with prospects, so as to elicit response from them. In a typical scenario, we often have a huge list of customers. But among the huge list, there are usually few real buyers which amount to a few percent [34]. Since the budget of a campaign is limited, it is important to focus the effort on the most promising prospects so that the response rate can be improved.

With the advancement of computing and database technology, people seek for computational approaches to assist in decision making. From the database that contains demographic details of customers, the objective is to develop a *response model* and use the model to predict promising prospects. The model needs to score each customer in the database with the likelihood of purchase. The customers are then ranked according to the score. A ranked list is desired because it allows decision makers to select the portion of customer list to roll out to [32]. For instance, out of the 200,000 customers on the list, we might wish to send out catalogs or brochures to the most promising 20% of customers so that the advertising campaign is cost-effective [33]. Hence, one way to evaluate the response model is to look at its performance at different *depth-of-file*. In the literature, there are various approaches proposed for building the response model. Here, we give a brief review in the following paragraphs.

In the recency-frequency-monetary model (RFM) [35], the profitability of a customer is estimated by three factors including the recency of buying, frequency of buying, and the amount of money one spent. Hence, only individuals that are profitable will be the targets of the campaign.

The Automatic Interaction Detection (AID) system uses tree analysis to divide consumers into different segments [35]. Later, the system was modified and became the Chi-Squared Automatic Interaction Detector (CHAID). The logistic regression model assumes that the logarithm of the *odd ratio* (*logit*) of the dependent variable (active or inactive respondents) is a linear function of the independent variables. The odd ratio is the ratio of the probabilities of the event happening to not happening. Because the approach is popular, newly proposed models are often compared with the logistic regression model as the baseline comparison [33, 36, 37].

Zahavi and Levin [37] examined the possibility of training a back-propagation neural network as the response model. However, due to a number of practical issues and that the empirical result did not improve over a logistic regression model, it seems that the neural network approach does not bring much benefit.

Ling and Li [38] combined the naïve Bayesian classifier and C4.5 to construct the response model. They evaluated their response model across three different real-life databases, the result illustrated that their approach are effective for solving the problem.

Bhattacharyya formulated the direct marketing problem as a multi-objective optimization problem [33, 36]. He suggested that the evaluation criterion should include the performance of the model at a given depth-of-file. In an early attempt [33], he used a genetic algorithm (GA) to learn the weights of a linear response model while the fitness evaluation function was a weighted average of the two evaluation criteria. When comparing the learnt model with the logit model on a real-life database, the new approach indicated a superior performance. Recently, he applied genetic programming (GP) to learn a tree-structured symbolic rule form as the response model [36]. Instead of using a weighted average criterion function, the new approach searches for *Pareto-optimal* solutions. From the analysis, he found that the GP approach outperforms the GA approach and is effective at obtaining solutions with different levels of trade-offs [36].

## 5.2 Methodology

The direct marketing database used here contains records of customers of a specialty catalog company, which mails catalogs to good customers on a regular basis. In this database, there are 5,740 active respondents and 14,260 non-respondents. The response rate is 28.7%. Each customer is described by 361 attributes. We selected nine attributes, which are relevant to the prediction, out of the 361 attributes. Missing values are then introduced randomly into the database. The percentages of the missing values in our experiments are 1%, 5%, and 10%, respectively.

We compare the performance of the Bayesian networks evolved by HEAm (HEAm models) with those obtained by LibB , Bayesware Discoverer (BayD), neural network (BNN) [37], logistic regression (LR), naïve Bayesian network classifier (NB) [39], and tree-augmented naive Bayesian network classifier (TAN) [39] from the database.

In the experiments, HEAm, LibB and BayD are executed directly on the incomplete database with missing values. For BNN, LR, NB, and TAN, the database is processed employing the mean value to replace missing values for each continuous variable and the mode for completing missing values for each discrete variable.

We use decile analysis here to compare the performance of different response models. It estimates the enhancement of the response rate for ranking at different depth-of-file. Essentially, the descending sorted ranking list is equally divided into 10 deciles. Customers in the first decile are the top ranked customers that are most likely to give response. On the other hand, customers in the tenth decile are the lowest ranked customers who are least likely to give response. Then, a *gains table* is constructed to describe the performance of the response model. In a gains table, we tabulate various statistics at each decile, including [40]:

- **Predicted Probability of Active:** the average of the predicted probabilities of active respondents in the decile by the response model.
- **Percentage of Active:** the percentage of active respondents in the decile.
- **Cumulative Percentage of Active:** the cumulative percentage of active respondents from decile 0 to this decile.
- **Actives:** the number of active respondents in this decile.
- **Percentage of Total Actives:** the ratio of the number of active respondents in this decile to the number of all active respondents in the database.
- **Cumulative Actives:** the number of active respondents from decile 0 to this decile.
- **Cumulative Percentage of Total Actives:** the ratio of the number of cumulative active respondents (from decile 0 to this decile) to the total number of active respondents in the database.
- **Lift:** It is calculated by dividing the percentage of active respondents by the response rate of the file. Intuitively, it estimates the enhancement by the respondence model in discriminating active respondents over a random approach for the current decile.
- **Cumulative Lift:** It is calculated by dividing the cumulative percentage of active respondents by the respondence rate. This measure evaluates how good the response model is for a given depth-of-file over a random approach. It provides an important estimate of the performance of the model.

### 5.3 Cross-Validation Results

In order to compare the robustness of the response models, we adopt a 10-fold cross-validation approach for performance estimation. The database is randomly partitioned into 10 mutually exclusive and exhaustive folds. For HEAm, LibB and BayD, the incomplete database is used. However, for the other methods, the

**Table 5.** Results of the networks evolved by HEAm for the database with 1% missing values

| Decile | Prob. of Active | % of Active | Cum. % of Active | Actives | % of Total Actives | Cum. Actives | Cum. % of Total Actives | Lift | Cum. Lift |
|---|---|---|---|---|---|---|---|---|---|
| 0 | 44.32% | 93.12% | 93.12% | 185.30 | 32.30% | 185.30 | 32.30% | 324.60 | 324.60 |
| | (0.35%) | (1.37%) | (1.37%) | (2.73) | (0.77%) | (2.73) | (0.77%) | (7.77) | (7.77) |
| 1 | 43.35% | 41.51% | 67.31% | 82.60 | 14.36% | 267.90 | 46.66% | 144.32 | 234.46 |
| | (0.25%) | (8.49%) | (4.61%) | (16.90) | (2.78%) | (18.35) | (2.74%) | (27.92) | (13.76) |
| 2 | 43.25% | 0.74% | 45.12% | 1.47 | 0.26% | 269.37 | 46.91% | 2.58 | 157.17 |
| | (0.22%) | (1.92%) | (2.66%) | (3.83) | (0.67%) | (15.88) | (2.25%) | (6.77) | (7.54) |
| 3 | 31.30% | 30.23% | 41.40% | 60.17 | 10.49% | 329.53 | 57.40% | 105.38 | 144.22 |
| | (1.58%) | (2.92%) | (1.84%) | (5.81) | (1.01%) | (14.62) | (1.84%) | (10.19) | (4.61) |
| 4 | 24.58% | 27.76% | 38.67% | 55.23 | 9.63% | 384.77 | 67.03% | 96.82 | 134.74 |
| | (0.35%) | (3.75%) | (1.36%) | (7.47) | (1.37%) | (13.54) | (1.73%) | (13.73) | (3.47) |
| 5 | 23.11% | 60.23% | 42.26% | 119.87 | 20.87% | 504.63 | 87.91% | 209.79 | 147.25 |
| | (0.24%) | (5.72%) | (1.40%) | (11.39) | (1.81%) | (16.74) | (1.56%) | (18.21) | (2.62) |
| 6 | 22.74% | 1.01% | 36.37% | 2.00 | 0.35% | 506.63 | 88.26% | 3.50 | 126.71 |
| | (0.15%) | (3.07%) | (1.17%) | (6.10) | (1.06%) | (16.28) | (1.41%) | (10.69) | (2.03) |
| 7 | 22.52% | 3.32% | 32.24% | 6.60 | 1.14% | 513.23 | 89.40% | 11.50 | 112.31 |
| | (0.33%) | (4.67%) | (1.11%) | (9.30) | (1.60%) | (17.62) | (1.51%) | (16.10) | (1.90) |
| 8 | 17.05% | 24.44% | 31.37% | 48.63 | 8.48% | 561.87 | 97.88% | 85.26 | 109.31 |
| | (0.37%) | (4.10%) | (0.82%) | (8.16) | (1.47%) | (14.71) | (0.55%) | (14.82) | (0.62) |
| 9 | 14.79% | 5.81% | 28.70% | 12.13 | 2.12% | 574.00 | 100.00% | 20.25 | 100.00 |
| | (0.41%) | (1.51%) | (0.71%) | (3.16) | (0.55%) | (14.17) | (0.00%) | (5.30) | (0.00) |
| Total | | | | 574.00 | | | | | |

**Table 6.** Results of the networks evolved by HEAm for the database with 5% missing values

| Decile | Prob. of Active | % of Active | Cum. % of Active | Actives | % of Total Actives | Cum. Actives | Cum. % of Total Actives | Lift | Cum. Lift |
|---|---|---|---|---|---|---|---|---|---|
| 0 | 45.60% | 88.78% | 88.78% | 176.67 | 30.77% | 176.67 | 30.77% | 309.27 | 309.27 |
| | (3.41%) | (12.06%) | (12.06%) | (24.01) | (4.11%) | (24.01) | (4.11%) | (41.34) | (41.34) |
| 1 | 43.25% | 35.54% | 62.16% | 70.73 | 12.31% | 247.40 | 43.08% | 123.74 | 216.51 |
| | (0.74%) | (7.02%) | (6.83%) | (13.98) | (2.36%) | (27.20) | (4.49%) | (23.75) | (22.58) |
| 2 | 41.87% | 8.04% | 44.12% | 16.00 | 2.79% | 263.40 | 45.88% | 28.08 | 153.70 |
| | (3.31%) | (12.52%) | (2.53%) | (24.92) | (4.38%) | (15.12) | (2.18%) | (44.04) | (7.31) |
| 3 | 30.55% | 31.32% | 40.92% | 62.33 | 10.86% | 325.73 | 56.74% | 109.17 | 142.57 |
| | (1.71%) | (2.28%) | (1.89%) | (4.54) | (0.79%) | (15.02) | (2.05%) | (7.91) | (5.14) |
| 4 | 24.57% | 32.41% | 39.22% | 64.50 | 11.25% | 390.23 | 67.99% | 113.02 | 136.66 |
| | (0.47%) | (5.91%) | (1.83%) | (11.76) | (2.04%) | (18.23) | (2.73%) | (20.53) | (5.49) |
| 5 | 23.45% | 50.30% | 41.07% | 100.10 | 17.41% | 490.33 | 85.39% | 174.96 | 143.04 |
| | (0.61%) | (20.95%) | (2.84%) | (41.70) | (7.19%) | (33.92) | (5.06%) | (72.27) | (8.47) |
| 6 | 22.72% | 4.04% | 35.78% | 8.03 | 1.42% | 498.37 | 86.81% | 14.22 | 124.64 |
| | (0.36%) | (7.72%) | (1.55%) | (15.36) | (2.71%) | (21.58) | (2.66%) | (27.26) | (3.82) |
| 7 | 22.29% | 6.13% | 32.07% | 12.20 | 2.12% | 510.57 | 88.93% | 21.33 | 111.72 |
| | (0.84%) | (10.96%) | (1.26%) | (21.80) | (3.75%) | (20.04) | (2.03%) | (37.73) | (2.55) |
| 8 | 17.37% | 25.85% | 31.38% | 51.43 | 8.98% | 562.00 | 97.91% | 90.26 | 109.34 |
| | (1.06%) | (6.01%) | (0.77%) | (11.95) | (2.11%) | (13.84) | (0.82%) | (21.23) | (0.91) |
| 9 | 15.23% | 5.74% | 28.70% | 12.00 | 2.09% | 574.00 | 100.00% | 19.97 | 100.00 |
| | (1.72%) | (2.28%) | (0.71%) | (4.76) | (0.82%) | (14.17%) | (0.00) | (7.80) | (0.00) |
| Total | | | | 574.00 | | | | | |

**Table 7.** Results of the networks evolved by HEAm for the database with 10% missing values

| Decile | Prob. of Active | % of Active | Cum. % of Active | Actives | % of Total Actives | Cum. Actives | Cum. % of Total Actives | Lift | Cum. Lift |
|---|---|---|---|---|---|---|---|---|---|
| 0 | 47.13% (4.74%) | 82.71% (17.43%) | 82.71% (17.43%) | 164.60 (34.69) | 28.66% (5.93%) | 164.60 (34.69) | 28.66% (5.93%) | 287.99 (59.63) | 287.99 (59.63) |
| 1 | 42.430% (1.62%) | 31.11% (9.92%) | 56.91% (7.40%) | 61.90 (19.75) | 10.77% (3.41%) | 226.50 (29.47) | 39.43% (4.81%) | 108.28 (34.28) | 198.14 (24.18) |
| 2 | 39.98% (4.30%) | 17.15% (13.09%) | 43.66% (2.10%) | 34.13 (26.04) | 5.97% (4.59%) | 260.63 (12.54) | 45.40% (1.78%) | 60.03 (46.18) | 152.10 (5.97) |
| 3 | 29.70% (1.59%) | 32.16% (8.34%) | 40.78% (2.25%) | 64.00 (16.60) | 11.16% (2.99%) | 324.63 (17.92) | 56.57% (3.05%) | 112.21 (30.06) | 142.13 (7.65) |
| 4 | 24.90% (0.60%) | 33.37% (10.48%) | 39.30% (1.46%) | 66.40 (20.85) | 11.58% (3.71%) | 391.03 (14.48) | 68.14% (2.50%) | 116.34 (37.34) | 136.97 (5.02) |
| 5 | 23.89% (0.97%) | 39.38% (26.34%) | 39.31% (4.11%) | 78.37 (52.41) | 13.56% (9.01%) | 469.40 (49.05) | 81.71% (7.55%) | 136.33 (90.50) | 136.86 (12.65) |
| 6 | 22.60% (0.58%) | 10.47% (12.30%) | 35.19% (2.03%) | 20.83 (24.48) | 3.67% (4.34%) | 490.23 (28.34) | 85.38% (3.89%) | 36.92 (43.58) | 122.58 (5.59) |
| 7 | 21.86% (1.31%) | 8.69% (10.76%) | 31.88% (1.27%) | 17.30 (21.42) | 3.02% (3.72%) | 507.53 (20.29) | 88.41% (2.22%) | 30.40 (37.41) | 111.06 (2.78) |
| 8 | 17.89% (1.33%) | 28.34% (8.85%) | 31.49% (0.92%) | 56.40 (17.62) | 9.84% (3.11%) | 563.93 (16.41) | 98.24% (1.21%) | 98.86 (31.30) | 109.71 (1.36) |
| 9 | 15.85% (2.03%) | 17.97% (3.27%) | 28.70% (0.71%) | 10.07 (6.84) | 1.76% (1.21%) | 574.00 (14.17) | 100.00% (0.00%) | 16.82 (11.62) | 100.00 (0.00) |
| Total | | | | 574.00 | | | | | |

**Table 8.** Cumulative lifts of the networks learned by different methods for the real-world databases with **1%** missing values

| Decile | HEAm | LibB | BayD | BNN | LR | NB | TAN |
|---|---|---|---|---|---|---|---|
| 0 | **324.60** (7.77) | $211.19^+$ (28.00) | $213.04^+$ (41.61) | $200.11^+$ (11.00) | $188.30^+$ (12.23) | $198.50^+$ (9.99) | $195.80^+$ (6.41) |
| 1 | **234.46** (13.76) | $185.59^+$ (17.44) | $189.43^+$ (14.53) | $171.01^+$ (9.76) | $168.80^+$ (9.73) | $169.70^+$ (7.15) | $168.30^+$ (7.35) |
| 2 | **157.17** (7.54) | 156.79 (7.08) | 155.99 (7.46) | 156.56 (5.74) | $152.30^+$ (6.72) | 154.30 (4.45) | $150.90^+$ (4.89) |
| 3 | 144.22 (4.61) | **146.54** (5.56) | 146.07 (7.90) | 144.26 (4.67) | $141.40^+$ (3.13) | $139.40^+$ (2.55) | $139.70^+$ (2.75) |
| 4 | 134.74 (3.47) | 136.43 (6.92) | **140.78** (12.08) | 135.60 (1.98) | $132.80^+$ (1.23) | $131.20^+$ (1.75) | 132.50 (4.17) |
| 5 | **147.25** (2.62) | $134.65^+$ (10.05) | $136.09^+$ (4.35) | $127.33^+$ (2.15) | $125.80^+$ (2.86) | $124.70^+$ (2.79) | $124.10^+$ (2.69) |
| 6 | **126.71** (2.03) | $119.16^+$ (4.11) | $119.63^+$ (1.82) | $120.20^+$ (2.02) | $118.30^+$ (2.26) | $116.70^+$ (1.64) | $118.70^+$ (1.70) |
| 7 | 112.31 (1.90) | 113.69 (3.87) | 112.53 (1.84) | $\mathbf{113.80}^-$ (1.61) | 112.50 (1.35) | 111.90 (1.45) | $113.40^-$ (1.17) |
| 8 | **109.31** (0.62) | 108.58 (2.03) | $107.64^+$ (1.86) | $107.71^+$ (0.98) | $106.60^+$ (1.07) | $106.20^+$ (0.92) | $106.20^+$ (1.03) |
| 9 | 100.00 (0.00) | 100.00 (0.00) | 100.00 (0.00) | 100.00 (0.00) | 100.00 (0.00) | 100.00 (0.00) | 100.00 (0.00) |

**Table 9.** Cumulative lifts of the networks learned by different methods for the real-world databases with **5%** missing values

| Decile | HEAm | LibB | BayD | BNN | LR | NB | TAN |
|---|---|---|---|---|---|---|---|
| 0 | **309.27** | $217.63^+$ | $246.59^+$ | $199.37^+$ | $188.50^+$ | $195.40^+$ | $197.80^+$ |
| | (41.34) | (47.64) | (31.34) | (10.33) | (11.45) | (10.27) | (9.84) |
| 1 | **216.51** | $186.30^+$ | $165.69^+$ | $171.09^+$ | $167.80^+$ | $170.30^+$ | $169.60^+$ |
| | (22.58) | (21.35) | (19.94) | (9.50) | (9.20) | (6.33) | (7.38) |
| 2 | 153.70 | 155.28 | 152.60 | **155.97** | 151.40 | 152.60 | 151.50 |
| | (7.31) | (6.96) | (7.80) | (5.60) | (4.77) | (4.14) | (5.23) |
| 3 | 142.57 | **145.15** | 143.24 | 143.21 | $140.40^+$ | $139.50^+$ | $139.90^+$ |
| | (5.14) | (8.33) | (6.71) | (3.67) | (2.67) | (2.72) | (2.85) |
| 4 | 136.66 | 136.75 | **144.16**$^-$ | 134.18 | $132.40^+$ | $130.50^+$ | $131.30^+$ |
| | (5.49) | (6.21) | (5.18) | (2.61) | (1.58) | (1.27) | (3.27) |
| 5 | **143.04** | $133.47^+$ | $124.27^+$ | $126.88^+$ | $125.60^+$ | $125.00^+$ | $123.60^+$ |
| | (8.47) | (10.49) | (3.38) | (2.49) | (2.67) | (2.62) | (1.65) |
| 6 | **124.64** | $118.90^+$ | $118.10^+$ | $120.07^+$ | $118.40^+$ | $117.00^+$ | $118.10^+$ |
| | (3.82) | (4.94) | (1.85) | (2.29) | (2.41) | (1.70) | (1.66) |
| 7 | 111.72 | $113.57^-$ | $113.09^-$ | **113.73**$^-$ | 112.40 | 111.50 | 112.50 |
| | (2.55) | (3.69) | (2.18) | (1.48) | (1.17) | (1.35) | (1.27) |
| 8 | **109.34** | $108.08^+$ | $106.80^+$ | $107.64^+$ | $106.60^+$ | $106.00^+$ | $106.10^+$ |
| | (0.91) | (1.89) | (1.56) | (0.87) | (0.97) | (1.15) | (1.10) |
| 9 | 100.00 | 100.00 | 100.00 | 100.00 | 100.00 | 100.00 | 100.00 |
| | (0.00) | (0.00) | (0.00) | (0.00) | (0.00) | (0.00) | (0.00) |

corresponding processed database is used. Each time, a different fold is chosen as the test set and other nine folds are combined as the training set. Bayesian networks are learned from the training set and evaluated on the corresponding test set.

In Table 5, the average of the statistics of the HEAm models for the database with 1% missing values at each decile are tabulated. Numbers in the parentheses are the standard deviations. The HEAm models have the cumulative lifts of 324.6 and 234.46 in the first two deciles respectively, suggesting that by mailing to the top two deciles alone, the Bayesian networks generate over twice as many respondents as a random mailing without a model. From Table 6, the evolved Bayesian networks for the database with 5% missing values achieve the cumulative lifts of 309.27 and 216.51 in the first two deciles. For the database with 10% missing values, the cumulative lifts in the first two deciles are 287.99 and 198.14, as shown in Table 7.

For the sake of comparison, the average of the cumulative lifts of the models learned by different methods from databases with different missing values are summarized in Table 8, Table 9, and Table 10, respectively. Numbers in the parentheses are the standard deviations. For each database, the highest cumulative lift in each decile is highlighted in bold. The superscript + represents that the cumulative lift of the HEAm models from the corresponding

**Table 10.** Cumulative lifts of the networks learned by different methods for the real-world databases with **10%** missing values

| Decile | HEAm | LibB | BayD | BNN | LR | NB | TAN |
|---|---|---|---|---|---|---|---|
| 0 | **287.99** | $239.06^{+}$ | $196.86^{+}$ | $195.71^{+}$ | $185.10^{+}$ | $190.40^{+}$ | $194.90^{+}$ |
| | (59.63) | (64.44) | (18.50) | (13.60) | (12.56) | (13.55) | (11.43) |
| 1 | **198.14** | $188.42^{+}$ | $171.22^{+}$ | $169.89^{+}$ | $164.90^{+}$ | $167.70^{+}$ | 167.20 |
| | (24.18) | (21.09) | (9.13) | (9.75) | (10.46) | (6.29) | (8.83) |
| 2 | 152.10 | 153.36 | 152.20 | **154.32** | 149.30 | 151.30 | 151.30 |
| | (5.97) | (6.38) | (6.40) | (6.76) | (8.11) | (3.95) | (5.38) |
| 3 | 142.13 | **142.46** | 139.63 | 142.28 | $138.90^{+}$ | $138.40^{+}$ | 139.40 |
| | (7.65) | (9.31) | (4.50) | (4.66) | (3.57) | (2.91) | (3.63) |
| 4 | **136.97** | 134.86 | $131.55^{+}$ | $133.14^{+}$ | $130.70^{+}$ | $128.60^{+}$ | $129.80^{+}$ |
| | (5.02) | (5.83) | (4.84) | (3.55) | (2.31) | (1.78) | (4.16) |
| 5 | **136.86** | $134.62^{+}$ | $124.17^{+}$ | $125.38^{+}$ | $123.60^{+}$ | $123.50^{+}$ | $123.20^{+}$ |
| | (12.65) | (10.86) | (5.17) | (1.82) | (2.01) | (1.72) | (1.99) |
| 6 | **122.58** | $119.65^{+}$ | $117.23^{+}$ | $119.27^{+}$ | $117.70^{+}$ | $116.10^{+}$ | $117.30^{+}$ |
| | (5.59) | (5.40) | (2.73) | (2.25) | (2.67) | (2.33) | (1.42) |
| 7 | 111.06 | **112.61** | 112.36 | $113.25^{-}$ | 111.90 | 111.20 | $112.50^{-}$ |
| | (2.78) | (4.21) | (1.85) | (1.28) | (1.85) | (1.81) | (1.27) |
| 8 | **109.71** | 108.97 | $105.51^{+}$ | $107.09^{+}$ | $106.40^{+}$ | $105.60^{+}$ | $106.30^{+}$ |
| | (1.36) | (1.81) | (1.22) | (0.67) | (0.84) | (0.97) | (0.82) |
| 9 | 100.00 | 100.00 | 100.00 | 100.00 | 100.00 | 100.00 | 100.00 |
| | (0.00) | (0.00) | (0.00) | (0.00) | (0.00) | (0.00) | (0.00) |

database is significant higher at 0.05 level than that of the models obtained by the corresponding methods. On the other hand, the superscript − represents that the cumulative lift of the HEAm models is significant lower at 0.05 level than that of the corresponding models.

In Table 8, the average and the standard deviations of the cumulative lifts of the models learned by different methods for the database with 1% missing values are shown. In the first two deciles, the networks learned by LibB have cumulative lifts of 211.19 and 185.59, respectively; and 213.04 and 189.43 for Bayeseware Discoverer models, respectively. It can be observed that HEAm models get the highest cumulative lifts in the first three deciles, and the cumulative lifts of the HEAm models in the first two deciles are significantly higher at 0.05 level than those of other models.

In Table 9, the average and the standard deviations of the cumulative lifts for different models learned from the database with 5% missing values are shown. In the first two deciles, the HEAm models have the highest cumulative lifts respectively, and they are significantly higher than those of corresponding methods at 0.05 level.

In Table 10, the average and the standard deviations of the cumulative lifts for different models discovered from the database with 10% missing values are shown. Again, it demonstrates that the discovered HEAm models have the highest cumulative lifts in the first two deciles respectively. The cumulative lifts of HEAm models in the first two deciles are significantly higher at 0.05 level than those of other eight corresponding methods.

To summarize, the networks generated by HEAm can always have the highest cumulative lifts in the first two deciles. Moreover, the cumulative lifts of the HEAm models are significantly higher at 0.05 level than those of other models in the first two deciles. We can conclude that HEAm is very effective in learning Bayesian networks from databases with different missing value percentages.

Since an advertising campaign often involves huge investment, a Bayesian network which can categorize more prospects into the target list is valuable as it will enhance the response rate. From the experimental results, it seems that HEAm are more effective than the other methods.

## 6 Conclusion

In this study, we propose a novel data mining system called HEAm that uses EM to handle incomplete databases with missing values and uses a hybrid evolutionary algorithm to search for good candidate Bayesian networks. The two procedures are iterated so that we can continue finding a better model while optimizing the parameters for a good model to complete the database with more accurate information. Instead of using the expected values of statistics as in most existing SEM algorithms, HEAm applies a data completing procedure to complete the database and thus decomposable scoring metrics can be used to evaluate the networks. Through comparison experiments on the databases generated from three benchmark networks structures, we demonstrate that HEAm outperforms LibB and Bayesware Discoverer.

We have also applied HEAm to a real-world direct marketing problem, which requires ranking the previous customers according to their probability of potential purchasing. The results show that the Bayesian networks obtained by HEAm outperform other models learned by other learning algorithms.

## Acknowledgements

This work is supported by the Lingnan University Direct Grant DR04B8.

## References

1. Jensen, F.V.: An Introduction to Bayesian Network. University of College London Press (1996)
2. Andreassen, S., Woldbye, M., Falck, B., Andersen, S.: MUNIN: A Causal Probabilistic Network for Interpretation of Electromyographic Findings. In: Proceedings of the Tenth International Joint Conference on Artificial Intelligence, pp. 366–372 (1987)
3. Cheeseman, P., Kelly, J., Self, M., Stutz, J., Taylor, W., Freeman, D.: AutoClass: a Bayesian classification system. In: Proceedings of the Fifth International Workshop on Machine Learning, pp. 54–64 (1988)

4. Heckerman, D., Horvitz, E.: Inferring Informational Goals from Free-Text Queries: A Bayesian Approach. In: Proceedings of the Fourteenth Conference on Uncertainty in Artificial Intelligence, pp. 230–237 (1998)
5. Heckerman, D., Wellman, M.P.: Bayesian Networks. Communications of the ACM 38(3), 27–30 (1995)
6. Cheng, J., Greiner, R., Kelly, J., Bell, D., Liu, W.: Learning Bayesian Networks from Data: An Information-Theory Based Approach. Artificial Intelligence 137, 43–90 (2002)
7. Spirtes, P., Glymour, C., Scheines, R.: Causation, Prediction, and Search. MIT Press, Cambridge (2000)
8. Cooper, G., Herskovits, E.: A Bayesian Method for the Induction of Probabilistic Networks from Data. Machine Learning 9(4), 309–347 (1992)
9. Heckerman, D.: A Tutorial on Learning Bayesian Networks. Tech. Rep. MSR-TR-95-06. Microsoft Research Adv. Technol. Div., Redmond, WA (1995)
10. Lam, W., Bacchus, F.: Learning Bayesian belief networks: an approach based on the MDL principle. Computational Intelligence 10(4), 269–293 (1994)
11. Larrañaga, P., Poza, M., Yurramendi, Y., Murga, R., Kuijpers, C.: Structure Learning of Bayesian Network by Genetic Algorithms: A Performance Analysis of Control Parameters. IEEE Transactions on Pattern Analysis and Machine Intelligence 18(9), 912–926 (1996)
12. Larrañaga, P., Kuijpers, C., Mura, R., Yurramendi, Y.: Learning Bayesian Network Structures by Searching for The Best Ordering with Genetic Algorithms. IEEE Transactions on System, Man and Cybernetics 26(4), 487–493 (1996)
13. Wong, M.L., Lam, W., Leung, K.S.: Using Evolutionary Programming and Minimum Description Length principle for data mining of Bayesian networks. IEEE Transactions on Pattern Analysis and Machine Intelligence 21(2), 174–178 (1999)
14. Wong, M.L., Leung, K.S.: An Efficient Data Mining Method for Learning Bayesian Networks Using an Evolutionary Algorithm-Based Hybrid Approach. IEEE Transactions on Evolutionary Computation 8(4), 378–404 (2004)
15. Schafer, J.L., Graham, J.W.: Missing data: Our View of the State of the Art. Psychological Methods 7(2), 147–177 (2002)
16. Ramoni, M., Sebastiani, P.: Efficient Parameter Learning in Bayesian Networks from Incomplete Databases. Tech. Rep. KMI-TR-41 (1997)
17. Ramoni, M., Sebastiani, P.: The Use of Exogenous Knowledge to Learn Bayesian Networks from Incomplete Databases. Tech. Rep. KMI-TR-44 (1997)
18. Friedman, N.: Learning Belief Networks in the Presence of Missing Values and Hidden Variables. In: Proceedings of the Fourteenth International Conference on Machine Learning, pp. 125–133 (1997)
19. Friedman, N.: The Bayesian Structural EM Algorithm. In: Proceedings of the Fourteenth Conference on Uncertainty in Artificial Intelligence, pp. 80–89 (1998)
20. Peña, J.M., Lozano, J.A., Larrañaga, P.: An Improved Bayesian Structural EM Algorithm for Learning Bayesian Networks for Clustering. Pattern Recognition Letters 21, 779–786 (2000)
21. Peña, J.M., Lozano, J.A., Larrañaga, P.: Learning Recursive Bayesian Multinets for Data Clustering by Means of Constructive Induction. Machine Learning 47, 63–89 (2002)
22. Myers, J., Laskey, K., DeJong, K.: Learning Bayesian Networks from Incomplete Data using Evolutionary Algorithms. In: Proceedings of the First Annual Conference on Genetic and Evolutionary Computation Conference, pp. 458–465 (1999)
23. Pearl, J.: Probabilistic Reasoning in Intelligent Systems: Networks of Plausible Inference. Morgan Kaufmann, San Mateo (1998)

24. Dempster, A.P., Laird, N.M., Rubin, D.B.: Maximum Likelihood from Incomplete Data via the EM Algorithm. Journal of the Royal Statistical Society(B) 39(1), 1–38 (1977)
25. Lauritzen, S.: The EM Algorithm for Graphical Association Models with Missing Data. Computational Statistics and Data Analysis 19, 191–201 (1995)
26. Huang, C., Darwiche, A.: Inference in Belief Networks: a Procedural Guide. International Journal of Approximate Reasoning 15(3), 225–263 (1996)
27. LibB, `http://compbio.cs.huji.ac.il/LibB/`
28. Bayesware Discoverer, `http://www.bayesware.com/frontpage.html`
29. Norsys Bayes Net Library, `http://www.norsys.com/net_library.htm`
30. Chickering, D.M.: Learning Equivalence Classes of Bayesian Network Structures. Journal of Machine Learning Research 2, 445–498 (2002)
31. Beaumont, G.P., Knowles, J.D.: Statistical Tests: An Introduction with MINITAB Commentary. Prentice-Hall, Englewood Cliffs (1996)
32. Zahavi, J., Levin, N.: Issues and Problems in Applying Neural Computing to Target Marketing. Journal of Direct Marketing 11(4), 63–75 (1997)
33. Bhattacharyya, S.: Direct Marketing Response Models using Genetic Algorithms. In: Proceedings of the Fourth International Conference on Knowledge Discovery and Data Mining, pp. 144–148 (1998)
34. Cabena, P., Hadjinian, P., Stadler, R., Verhees, J., Zanasi, A.: Discovering Data Mining: From Concept to Implementation. Prentice-Hall, Englewood Cliffs (1997)
35. Petrison, L.A., Blattberg, R.C., Wang, P.: Database Marketing: Past, Present, and Future. Journal of Direct Marketing 11(4), 109–125 (1997)
36. Bhattacharyya, S.: Evolutionary Algorithms in Data Mining: Multi-Objective Performance Modeling for Direct Marketing. In: Proceedings of the Sixth International Conference on Knowledge Discovery and Data Mining, pp. 465–473 (2000)
37. Zahavi, J., Levin, N.: Applying Neural Computing to Target Marketing. Journal of Direct Marketing 11(4), 76–93 (1997)
38. Ling, C.X., Li, C.H.: Data Mining for Direct Marketing: Problems and Solutions. In: Proceedings of the Fourth International Conference on Knowledge Discovery and Data Mining, pp. 73–79 (1998)
39. Friedman, N., Geiger, D., Goldszmidt, M.: Bayesian Network Classifiers. Machine Learning 29, 131–163 (1997)
40. Rud, O.P.: Data Mining Cookbook: Modeling Data for Marketing, Risk and Customer Relationship Management. Wiley, New York (2001)

# Fuzzy Local Currency Based on Social Network Analysis for Promoting Community Businesses

Osamu Katai, Hiroshi Kawakami, and Takayuki Shiose

Graduate School of Informatics, Kyoto University, Sakyo, Kyoto 606-8501, Japan
{katai,kawakami,shiose}@i.kyoto-u.ac.jp

**Summary.** This paper discusses the ability of local currencies (LCs) to exchange goods and/or services by introducing a method to analyze the reciprocity of communities based on fuzzy network analysis. LCs are expected to revitalize social communities that face difficulties due to the attenuation of human relations. Therefore, such currencies have drastically spread all over the world to resolve these difficulties. LCs circulate in particular areas or communities and enhance social capitals. The significance of reciprocity in a community is usually referred to in light of the non-additivity of evaluation measures that reflect the non-additivity of relationships among community members and/or their activities. To analyze such reciprocity, we employ a fuzzy measure based on fuzzy network analysis that provides certain guidelines for the emergence of interpersonal relationalities among community members.

**Keywords:** Local currency, Reciprocity, Fuzzy network analysis, Fuzzy measure, Choquet integral.

## 1 Introduction

Communities, which are essential for daily mutual aid and social activities, are expected to get greater abilities for enhancing such activities. Nevertheless, communities seem to be declining recently due to the attenuation of human relations. Local currencies are becoming popular in the world for resolving this problem. We expect that our communities will be more lively and harmonious by using these currencies. The main feature of local currencies, different from national global currencies, reflects reciprocity to enhance social relationships through mutual aid in communities.

Using fuzzy logic, this paper proposes a method for evaluating the reciprocity of local currencies, and discusses how to reflect the emergence of social capital through exchanging goods and/or services by local currencies among community members.

Section 2 introduces the notion of local currencies with examples and discusses expected properties, i.e., reciprocity. Based on the fuzzy network analysis introduced in Section 3, Section 4 proposes an evaluation method of reciprocity in a community. Finally, we discuss the proposed method as well as how to get the guidelines for the emergence of meaningful interpersonal relationships in a community in Section 5.

M. Gen et al.: Intelligent and Evolutionary Systems, SCI 187, pp. 37–48.
springerlink.com 

# 2 Local Currency

## 2.1 General Perspectives

A local currency is defined as one that circulates within a local area or inside a social group. As usual currencies, people use a local currency to exchange goods or services and to communicate with each other in a community whose members trust each another. Local currencies, which have spread all over the world [1], are classified into three types with respect to their purposes: to promote local economies, to support mutual aid, and a combination of these two. This paper focuses on the second type: the role of supporting mutual aid.

## 2.2 Essence of Local Currencies

According to Niklas Luhmann, a currency has bilateral characteristics called "symbolic" and "diabolic," that serve opposite effects, i.e., linking and separating people [2]. But it is impossible to separate these two characteristics from each other. Luhmann added that a currency is a medium that symbolically emerges through generalization. Generally, symbolic generalizations are made with three aspects (dimensions): temporal, eventual, and social. This means that currencies can be used whenever, for whatever, and with whomever. Generalizations provide currencies with their basic functions: storing values, measuring values, and functioning as a medium of exchange. Therefore they are considered communication media that provide opportunities to communicate with each other. On the other hand, the diabolic character of currency relates to the diabolic aspect of symbolic generalization. For instance, diabolic character leads to financial crisis, the supremacy of money, and economic disparity, etc. Luhmann argued that the most diabolic character is the attenuation of reciprocity.

Local currencies partly restrain the symbolic generalization to limit the diabolic aspects and re-link people. In other words, national currencies are communication media that discourage reciprocity while local currencies are communication media that encourage reciprocity.

## 2.3 Reciprocity

This paper interprets reciprocity as a general tendency on mutual exchanges in a community. A person feels pressure to contribute to restoring the community balance in the long term, even though all community members may feel an imbalance at each instant of time [3]. A payment with national currencies culminates in exchanges. Reciprocal exchanges with local currency, on the contrary, sustain community exchange for balance. Furthermore, reciprocal exchanges may be considered gifts.

One traditional example of a reciprocal exchange system is the "Kula-Ring." In the Trobriand Islands in the southeast corner of Papua, New Guinea, the Kula Ring is a unique and fascinating circular transaction system with two shell ornaments within the Trobriand society. A transaction involves transferring an ornament clockwise through the island network, while another ornament moves counterclockwise. Thus, this system forms a huge circle. The objective of the Kula Ring is not merely economical gains but also to reinforce interpersonal relationships.

## 2.4 Time Dollar

One modern example that reflects reciprocity is the time dollar[1]. The regions adopting this system have spread all over the United States and involve more than 200 projects. The original concept of the time dollar proposed by Edgar S. Cahn in 1980 is as follows. People who want to join the time-dollar system must register with the secretariat (coordinator), who regularly publishes a journal through which people can get information on the goods and services offered or requested by members. Then, a registered member may contact another member through an introduction by the secretariat. The essential characteristic of the time dollar is pricing the unit of the currency, i.e., an hour. This means that whoever a person is and whatever the service or good is, if it takes an hour, then it is worth one time-dollar. People may feel an imbalance of dealing each time, but time dollar focuses on the balance of dealing in the long term. Moreover, remarkably, time dollars cannot be exchanged with the national currency and are interest free, that is, a zero interest rate. Hence, there is no duty to repay because their purpose is to support gifts; there is no meaning to save them. This time-dollar system is based on trust among members and is quite different from national currencies which we are accustomed. People who join the time-dollar system are interested in not only receiving benefits or convenience but also contributing to their community and helping each other.

In the next subsection, we introduce social capital, which is the final objective of the time-dollar system of Cahn. It provides us with standpoints for considering the roles of reciprocity in a community.

## 2.5 Social Capital

The notion of social capital provides a useful way of discussing civil society. Social capital expresses three basic social features: social networks, the norms of reciprocity, and trustworthiness, all of which enable efficient community collaboration to pursue our common purposes [4]. A significant property of social capital is how the three features affect each other (Fig. 1). If they are strengthened then a community is revitalized in a virtuous circle; if weakened, then a community declines in a vicious circle. Thus when trustworthiness is weakened in a community, it is difficult to strengthen it from the beginning, but it is possible to strengthen social networks and the norms of reciprocity. From this viewpoint, it is possible to determine the validity of a local currency.

In the analysis of social capital in real societies, W. Baker and J. Kanamitsu evaluated it based on network analysis [5, 6]. But since they disregarded

[1]http://www.timedollar.org/index.htm

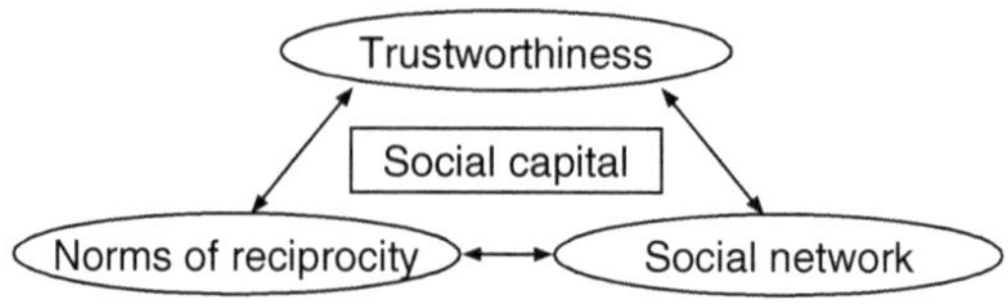


**Fig. 1.** Concept of social capital

reciprocity in a community, their research did not treat the social capital examined by R. D. Putnam. Therefore, we propose an evaluation method of reciprocity based on fuzzy network analysis.

# 3 Fuzzy Network Analysis

## 3.1 Fuzzy Graphs

For network analysis, we often use graph theory, even though two-valued logic is inadequate to address various problems in real societies. Thus the notion of a fuzzy graph is suitable to deal with the many-valuedness of real societies and to carry out mathematical analysis [7].

*[Def. of Fuzzy Graph]:* Let $N$ be a finite set, let $\tilde{N}$ be a fuzzy set over a set $N$ (the universe of discourse), and let $\tilde{L}$ be a fuzzy set s.t. $\tilde{L} \in F(N \times N)$. If the following holds:

$$\tilde{L}(x_i, x_j) \leq \tilde{N}(x_i) \wedge \tilde{N}(x_j)$$

for $\forall x_i, \forall x_j \in N$, then $G = (\tilde{N}, \tilde{L})$ is said to be a fuzzy graph. Connections between nodes $i$ and $j$ in the fuzzy graph are defined as follows:

$$r_{ij} : \begin{cases} 0 < r_{ij} \leq 1, & \text{if nodes } i \text{ and } j \text{ are connected} \\ r_{ij} = 0, & \text{if nodes } i \text{ and } j \text{ are disconnected.} \end{cases}$$

The relation of the connection in the fuzzy graph is considered a fuzzy relation over $N$. Let the cardinal number of $N$ be $n$, and then the relation of the connection is given as the fuzzy matrix:

$$R = (r_{ij})_{n \times n},$$

where $R$ is called the fuzzy adjacency matrix. Note that any fuzzy adjacency matrix $R$ is reflexive, i.e., $r_{ii} = 1$ for $\forall i$.

## 3.2 $\alpha$-cut

*[Def. of $\alpha$-cut]:* Let $\tilde{A}$ be a fuzzy set whose membership grade is given by $\lambda_{\tilde{A}}$ and $\alpha \in [0, 1]$. Then the crisp set is given as:

$$(\tilde{A})_\alpha = \{u \mid \lambda_{\tilde{A}}(u) > \alpha, u \in U\},$$

which is called the (strong) $\alpha$-cut of fuzzy set $\tilde{A}$, where $U$ is a universal set. Similarly, let $R$ be a fuzzy adjacency matrix. Matrix $(R)_\alpha = (r_{ij}^\alpha)_{n\times n}$ is called the $\alpha$-cut of $R$, where

$$r_{ij}^\alpha = \begin{cases} 1 \text{ if } r_{ij} > \alpha \\ 0 \text{ if } r_{ij} \le \alpha. \end{cases}$$

### 3.3 Fuzzy Measure

*[Def. of Fuzzy Measure]:* Let $(X, \mathcal{F})$ be a measurable space. If $\mu$: $\mathcal{F} \to [0, \infty]$ is defined as:

$$\mu(\emptyset) = 0,$$
$$A, B \in \mathcal{F}, A \subset B \Rightarrow \mu(A) \le \mu(B),$$

then $\mu$ is called a fuzzy measure over $\mathcal{F}$. Here, the triple $(X, \mathcal{F}, \mu)$ is called a fuzzy measure space. Conventional measures, e.g., probability measures, are specialized fuzzy measures satisfying the following additivity of measures:

$$A \cap B = \emptyset \Rightarrow \mu(A \cup B) = \mu(A) + \mu(B).$$

Generally, fuzzy measures do not presume the above additivity. Due to the lack of the above additivity, we have the following three cases with which the corresponding interpretations on the underlying social structures are associated:

case 1: $\mu(A \cup B) > \mu(A) + \mu(B)$: A positive (enhancing) synergy effect exists between events (or groups) $A$ and $B$.
case 2: $\mu(A \cup B) < \mu(A) + \mu(B)$: A negative (inhibitory) synergy effect exists between $A$ and $B$.
case 3: $\mu(A \cup B) = \mu(A) + \mu(B)$: $A$ and $B$ are independent of each other.

Thus fuzzy measures naturally reflect the effects of internal interactions among groups or systems by their essential characteristics, i.e., the non-additivity of measures.

### 3.4 Choquet Integral

We briefly introduce the Choquet integral, defined over non-additive measures [8].

*[Def. of Choquet Integral]:* For the following stepwise function

$$f(x) = \sum_{i=1}^{n} r_i 1_{D_i}(x),$$

where $0 < r_1 < r_2 < \cdots < r_n$, $D_i \cap D_j = \emptyset$ for $i \neq j$, and $1_{D_i}$ is the characteristic function of $D_i$, the Choquet integral of $f$ w.r.t. $\mu$ is defined as follows:

$$(C) \int f \, d\mu = \sum_{i=1}^{n} (r_i - r_{i-1}) \mu(A_i),$$

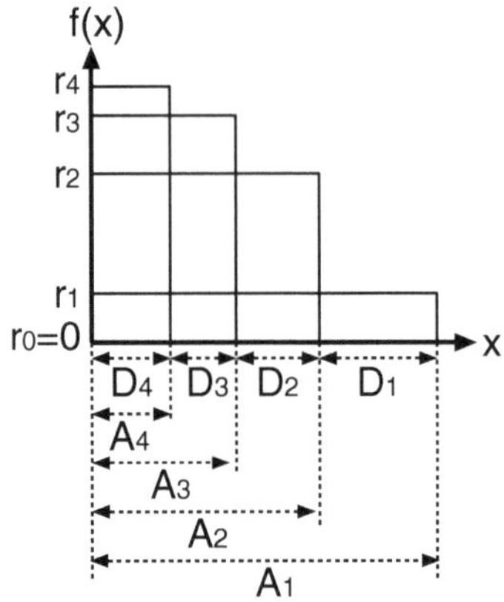


**Fig. 2.** Stepwise function integrated with value $r_i$ in domain $D_i$ for $i = 1, 2, 3, 4$

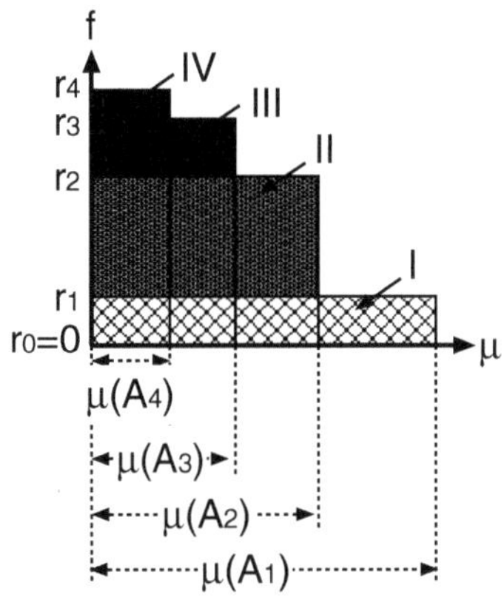


**Fig. 3.** Choquet integral of stepwise function as summation of horizontal columns I, II, III, and IV

where $r_0 = 0$ and $A_i = \cup_{j=i}^{n} D_j$. For example, when $n = 4$, the stepwise function is written as (cf. Fig. 2):

$$f(x) = \sum_{i=1}^{n} r_i 1_{D_i}(x) = \sum_{i=1}^{4} (r_i - r_{i-1}) 1_{A_i}(x).$$

Thus, the Choquet integral of $f$ w.r.t. $\mu$ is represented as (cf. Fig. 3):

$$(C) \int f \, d\mu = I + II + III + IV,$$

$$\begin{aligned} I &= (r_1 - r_0) \cdot \mu(A1), \\ II &= (r_2 - r_1) \cdot \mu(A2), \\ III &= (r_3 - r_2) \cdot \mu(A3), \\ IV &= (r_4 - r_3) \cdot \mu(A4). \end{aligned}$$

## 4 Analysis of Reciprocity

In this section we propose a novel evaluation framework to deal with a local currency in a community. For reciprocity in a community, the flow of goods or services is important because reciprocity is inseparable from the phrenic load of the gifts of others for three obligations: giving, receiving, and repaying [9]. Reciprocity must get greater values to evenly maintain the balance of the flow. Therefore, the value of reciprocity in the group is not the summation of individual transactions due to the non-additivity of the measure of groups.

### 4.1 Fuzzy Adjacency and Reachability of Community

First, we draw a fuzzy graph of a community, where nodes denote the members of a group and links reflect the amount of trade between them. Let $t_{ij}$ be the evaluation of the amount of services provided by member $i$ toward member $j$. Then the trade matrix is defined as

$$T = (t_{ij})_{n\times n},$$

where $n$ is the number of members in the group. We set

$$max\_income = \sup_i (\sum_j t_{ij})$$

for evaluating the maximum amount of received local currency. Next, we introduce the following fuzzy adjacency matrix

$$R = (r_{ij})_{n\times n} = T/max\_income$$

with the following properties:

$$r_{ij} \geq 0 \quad \text{for} \quad 1 \leq i \leq n, 1 \leq j \leq n,$$
$$\sum_j r_{ij} \leq 1 \quad \text{for} \quad 1 \leq i \leq n.$$

Then we introduce the reachability matrix $M$ of $R$ as follows:

$$M = I \oplus R \oplus R^2 \oplus \cdots \oplus R^{n-1},$$

where $\oplus$ is the bounded sum operation defined as $a \oplus b = 1 \wedge (a + b)$ and $I$ is the unit matrix of $n \times n$.

This definition is based on the concept of currency called Propagation Investment Currency SYstem (PICSY) [10]. Suppose that member $a$ provides $b$ a service whose evaluation by $b$ is 0.3, and member $b$ provides $c$ a service whose evaluation by $c$ is 0.4 (Fig. 4). Then, in PICSY, $a$ also receives evaluation by $c$ whose amount is the product $0.3 \times 0.4 = 0.12$.

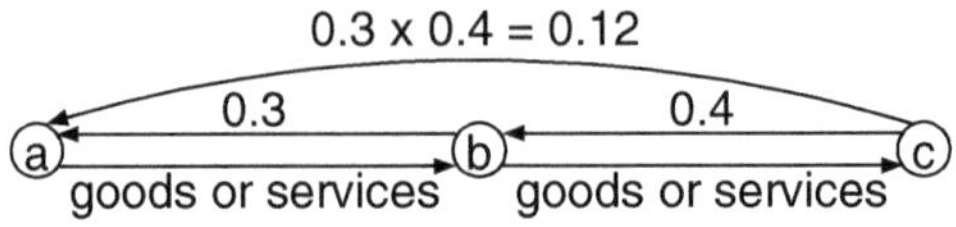


**Fig. 4.** Propagation Investment Currency SYstem (PICSY) concept

Let us examine the case with the following trade matrix:

$$T = \begin{pmatrix} 0\,3\,0\,3 \\ 0\,0\,5\,5 \\ 0\,8\,0\,0 \\ 0\,2\,0\,0 \end{pmatrix}.$$

Fuzzy adjacency matrix $R$ is calculated as

$$R = \frac{1}{10}\begin{pmatrix} 0\,3\,0\,3 \\ 0\,0\,5\,5 \\ 0\,8\,0\,0 \\ 0\,2\,0\,0 \end{pmatrix} = \begin{pmatrix} 0 & 0.3 & 0 & 0.3 \\ 0 & 0 & 0.5 & 0.5 \\ 0 & 0.8 & 0 & 0 \\ 0 & 0.2 & 0 & 0 \end{pmatrix}.$$

We finally obtain the reachability matrix as follows:

$$M = \begin{pmatrix} 1 & 0.69 & 0.27 & 0.57 \\ 0 & 1 & 0.75 & 0.75 \\ 0 & 0.3 & 1 & 0.1 \\ 0 & 0.1 & 0.4 & 1 \end{pmatrix}.$$

### 4.2 $\alpha$-cut and Structural Interpretation of Community

Next, let us consider the levels of reciprocity because reciprocity changes based on the social distance within a community [?]. For instance, a low level of reciprocity takes the form of greetings, while a high level of reciprocity takes the

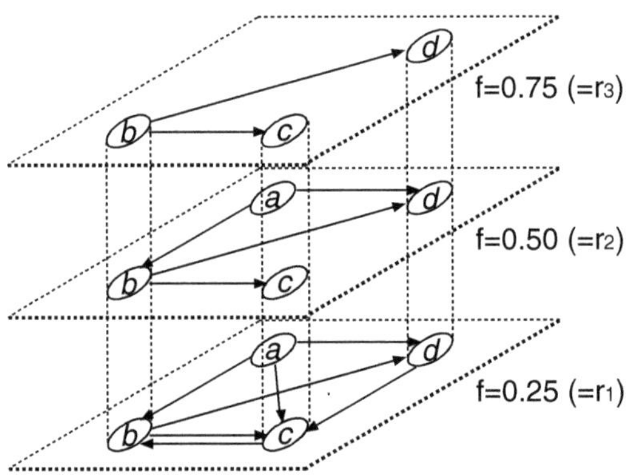


**Fig. 5.** $\alpha$-cut structurization of community

form of actions for local revitalization. For this purpose, we introduce $\alpha$-cut by $f$ for this adjacency matrix that also elucidates its structure. For example, if we set levels $f$ as 0.25, 0.5, and 0.75, we obtain

$$(M)_{0.25} = \begin{pmatrix} 1\,1\,1\,1 \\ 0\,1\,1\,1 \\ 0\,1\,1\,0 \\ 0\,0\,1\,1 \end{pmatrix}, \quad (M)_{0.5} = \begin{pmatrix} 1\,1\,0\,1 \\ 0\,1\,1\,1 \\ 0\,0\,1\,0 \\ 0\,0\,0\,1 \end{pmatrix}, \quad (M)_{0.75} = \begin{pmatrix} 1\,0\,0\,0 \\ 0\,1\,1\,1 \\ 0\,0\,1\,0 \\ 0\,0\,0\,1 \end{pmatrix},$$

respectively. The $\alpha$-cut structure consisting of these three cases is illustrated in Fig. 5

### 4.3 Reciprocity Analysis of Community

Next we evaluate reciprocity, which we consider the balance of the phrenic load of others' gifts; notice two measures, integration ($I(j)$) and radiality ($R(j)$) of member $j$. $I(j)$ indicates the degree to which individual $j$ is connected and $R(j)$ reflects the degree of reachability within a network. $I(j)$ is based on inward ties, and $R(j)$ is based on outward ties [12]. $I(j)$ and $R(j)$ are interpreted as the degree of benefits from a community and the degree of contributions to community.

*[Def. of Integration Measure]:* Let $D(= (d_{ij})_{n\times n})$ be a distance matrix and $n$ be the number of nodes. Then integration measure $I(j)$ for node $j$ is defined as:

$$I(j) = \frac{\sum_{i\neq j} \tilde{d}_{ij}}{n-1},$$

where $\tilde{d}_{ij}$ is called reverse distance, given as:

$$\tilde{d}_{ij} = \text{diameter} - d_{ij} + 1,$$

where the diameter is given as the maximum value within the distance matrix. The lower the value of a distance is, the higher the value of its reverse distance is.

*[Def. of Radiality Measure]:* Similarly, radiality measure $R(j)$ for node $j$ is defined as:

$$R(j) = \frac{\sum_{i\neq j} \tilde{d}_{ji}}{n-1}.$$

We propose a reciprocity measure on a fuzzy network based on integration and radiality measures. For this purpose, reverse distance $\tilde{D}$ $(= (\tilde{d}_{ij})_{n\times n})$ in a fuzzy network is modified with $\alpha$-cut by $f$ as:

$$\tilde{D}_f = M \wedge (M)_f.$$

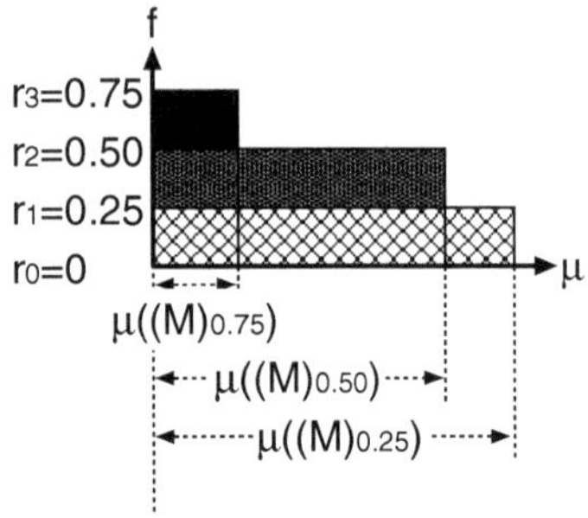


**Fig. 6.** Reciprocity in a community evaluated by Choquet Integral

*[Def. of Reciprocity Measure]:* The reciprocity measure of an individual in a fuzzy network with $\alpha$-cut by $f$ is defined as:

$$\mu_{(individual)}(j_f) = \frac{I_f(j) + R_f(j)}{(2 + |I_f(j) - R_f(j)|)},$$

where $I_f$ and $R_f$ are calculated by substituting $(\tilde{d}_{ij})_f$ into $\tilde{d}_{ij}$ in the definitions of $I(j)$ and $R(j)$.

Reciprocity gets a high score when both the integration and radiality measures are high and their difference is small. The reciprocity measure for the network with $\alpha$-cut by $f$ is also operationally defined as:

$$\mu((M)_f) = \sum_{j=1}^{n} \frac{I_f(j) + R_f(j)}{(2 + |I_f(j) - R_f(j)|)}.$$

This value represents the degree of the network, that is, being reciprocally connected. Reciprocity is represented as the sum of rectangular blocks described in Fig. 6:

$$\text{Reciprocity of } R = (C)\int f\, d\mu = \sum_{i=1}^{n}(r_i - r_{i-1})\mu((M)_f).$$

For the group with a three-levels cut shown in Fig. 6, the reciprocity of the group is calculated as:

$$\begin{aligned}
&\text{Reciprocity of } R \\
&= 0.25 \cdot \mu((M)_{0.25}) + (0.5 - 0.25) \cdot \mu((M)_{0.5}) + (0.75 - 0.5) \cdot \mu((M)_{0.75}) \\
&= 0.25 \times 1.430 + 0.25 \times 1.167 + 0.25 \times 0.286 \\
&= 0.721.
\end{aligned}$$

### 4.4 Community Business Orientation

This reciprocity measure can be used to decide the direction of business activities and service promotions. Suppose that member $b$ of the above group wants to

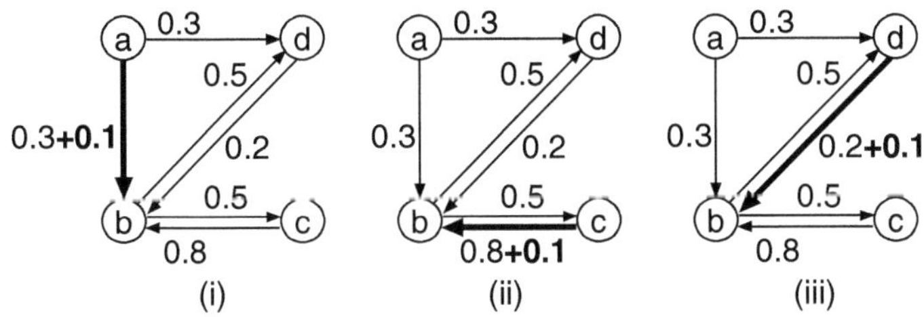


**Fig. 7.** Three cases of providing service toward $b$ from $a$, $c$, or $d$

receive a service with amount 1. Then incremental amount $\Delta r$ of the adjacency matrix is given as

$$\Delta r = \frac{1}{max\_income} = 0.1.$$

Thus we have the three possible adjacency matrices shown in Fig. 7 whose reciprocity values are:

case (i): 0.797, case (ii): 0.866, and case (iii): 0.868.

This result implies that member $b$ should receive the service from member $c$ or $d$.

## 5 Discussion and Conclusion

In this paper, we introduced a design concept of fuzzy local currency for constructing lively communities and for considering the reciprocity that can be expected to coordinate with the usage of the local currency. Reciprocity contributes to emerge and accumulate social capital. Thus we proposed an evaluation method for reciprocity using a fuzzy network analysis of a social community. Note that we can calculate the parameters in this analysis, despite the non-additive nature of the evaluation measure. The non-additivity of the evaluation measures reflects the non-additive relationships among community members or their activities. Furthermore, the secretariat of the local currency can obtain useful suggestions from this evaluation method. For example, in the case of Fig. 7, it can be readily seen that member $b$ should receive more goods or services from members $c$ or $d$ rather than from member $a$ to construct a lively community, because transactions from members $a$, $c$, or $d$ to member $b$ increase the reciprocity measure, as shown above. Using this information, the secretariat can promote transactions with local currency effectively. The community secretariat should not just wait for the outcome of the local currency but should promote its circulation to moderately control the emergence of social capital.

## References

1. Lietaer, B.A.: Das Geld Der Zukunef (1999); translated by Kobayashi, K., Hukumoto, H., Kato, S.: Collapse of money. Nihon Keizai Hyoronsha, Japan (in Japanese) (2000)

2. Luhmann, N.: Die Wirtschaft der Gesellschaft. Suhrkamp Verlag, Frankfurt (1988); translated by Kasuga, J.: Economics of Society. Bunshindo, Japan (in Japanese) (1991)
3. Konma, T.: The social anthropology of gift and exchange. Ochanomizu Syobo, Japan (in Japanese) (2000)
4. Putnam, R.D.: Bowling alone: The collapse and revival of American community. Simon Schuster, New York (2000)
5. Baker, W.: Achieving success through social capital. Jossey-Bass, San Francisco (2000)
6. Kanamitsu, J.: The base of social network analysis. Keisousyobou, Japan (in Japanese) (2003)
7. Ka, I., Oh, K.: Fuzzy network engineer. Nihon Rikou Syuppankai, Japan (in Japanese) (1995)
8. Grabisch, M., Murofushi, T., Sugeno, M.: Fuzzy measure and integral: theory and applications. Physica Verlag, Heidelberg (2000)
9. Mauss, M.: The gift. Cohen West, London (1954)
10. Nishibe, T.: The frontier of evolutionary economics. Nippon-Hyoron-Sha Co., Ltd., Japan (in Japanese) (2004)
11. Sahlins, M.: Stone age economics. Aldine, New York (1972)
12. Valente, T.W.: Integration and radiality: Measuring the extent of an individual's connectedness and reachability in a network. Social network 20(1), 89–105 (1998)

# Evolving Failure Resilience in Scale-Free Networks

George Leu and Akira Namatame

Dept. of Computer Science, National Defense Academy
leugeorge@yahoo.com, nama@nda.ac.jp

**Summary.** Today our society tends to become more and more dependent on large scale (global) infrastructure networks. In many cases, attacks on a few important nodes of such systems lead to irreparable local or, worse, global damages. Thus, designing resilient networks rather than reducing the effects of some unexpected attacks becomes a must. As the most resilient network, regarding any kind of attacks, should be a full-connected graph, it is obvious that implementing such a network is a utopia. This paper proposes an original multi-objective method for optimizing complex networks' structure, taking into account the implementation costs. A micro genetic algorithm is used in order to improve networks' resilience to targeted attacks on HUB nodes while keeping the implementation costs as low as possible.

## 1 Introduction

Most of the existing complex networks, such as internet, power transmission grids, world-wide terrestrial, maritime or air transportation networks, are believed to have a similar statistical characteristic, power law distribution of the nodes degrees; they are so called scale-free networks. From the connectivity point of view, scale-freeness provides a well known tolerance to random failures, but they are susceptible to failures of the high-connected (HUB) nodes. Attacks on these specific nodes maylead to a very fast disintegration of the whole network.

In the last years, a huge effort has been done to analyze this vulnerability, in order to improve networks' structure and their resilience, respective. Usually, the optimized networks have been found by trying to find new analytical representations, with best results obtained after many analysis and great amount of calculation and time [4,5,7,19].

Recently, new heuristic methods based on the Genetic Algorithms (GAs) have been used to optimize networks' structure, taking in account simple attributes, such as number of links/nodes and their connectivity, degree distribution, degree sequence [1,9].

In this paper, an original method for designing low-cost networks resilient to targeted attacks is proposed. The aim is to obtain a network having the lowest implementation cost and the highest resilience to targeted attacks or, depending on the designer's goal, specific weights for the two objectives can be assigned in order to improve the performance in the desired way.

## 2 Multi-objective Optimization Problem

As the aim is to improve the network's resilience and keep the cost at a low level, it is obvious that a multi-objective problem has to be solved. Thus an objective function

M. Gen et al.: Intelligent and Evolutionary Systems, SCI 187, pp. 49–59.
springerlink.com 

has to be found. Apart from the complex and time consuming analytical approach, the objective function shows itself in a very natural way, by simply looking at the network while thinking of the two opposite goals. In other words, it is enough to "picture" the ideas of *resilience* and *low cost* and the desired function yields clearly as sum of the two objective functions.

**The first objective function** is related to resilience and has to measure and improve the ability of the network to remain connected if attacked. Maximizing this function will conduct to a very robust graph regarding attacks on high-connected nodes. The strength of the network can be measured using *topological integrity* defined as follows:

$$G = \frac{N'}{N} \tag{1}$$

where $N'$ is the number of nodes in the largest connected network after the attack and $N$ is the initial number of nodes. Assuming that the network is continuously attacked, nodes being removed one by one starting from most to less connected nodes, the picture of this attack and the network's integrity will look like in Fig. 1

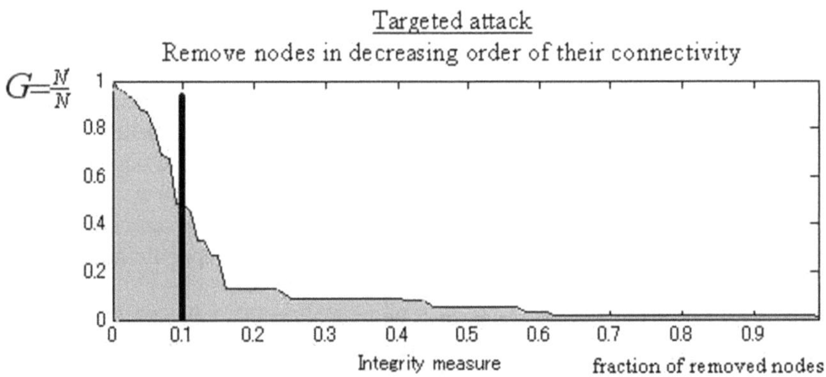


**Fig. 1.** Topological integrity of the network. For a Scale-Free network only 10% most connected nodes are taken in account. The other nodes are low connected, being irrelevant for the optimization.

Trying to improve the robustness of a network using this picture is the same thing with maximizing the hashed area situated under the graph of *G*. Still, as the network that is to be optimized is a Scale-Free network, only a few nodes are very high connected, most of them being low and very low connected. Thus, only first 10% important nodes are taken into account for the aria maximization, rest of the nodes being irrelevant for the optimization process. According to this way of thinking, the first objective function will be like in Eq. (2)

$$f_{obj1} = \sum_{upto10\%} G = \sum_{upto10\%} \frac{N'}{N} \tag{2}$$

Note that the best network from the first objective point of view will always be a full (or almost full) connected graph impossible to be implemented in the real world, mainly because of the costs, but not only. For this reason, a constraint in the number of links is needed, the second objective that is.

**The second objective function** is related to implementation costs. "Cost" in general is a very wide notion. It consists of and depends on many other notions, such as traffic, energy, distances, clients, goods etc. If talked about cost in general then it was very difficult to find a way for minimizing it without taking in account everything, an enormous waste of processing power and time. For this reason a simple way of defining costs has been imagined, as follows. If one is trying to find an objective function for minimizing costs by simply looking at the network structure, then a basic definition of cost should only include the number of edges that graph has. Basically speaking number of edges is indeed proportional with the implementation cost and no other parameter is needed when the topology only is to be optimized. Of course, the subject of the present paper could be anytime enlarged and many other things could be taken into account, but in this scenario only physical connectivity (no traffic, no distance) is relevant for the optimization process. According to this way of thinking the second objective will be like in Eq. (3)

$$f_{obj2} = E/E_{\max} \quad (3)$$

where $E$ is number of edges of the obtained graph and $E_{max}$ is the maximum number of edges (of the full-connected graph). The number of edges has been normalized to the number of edges of the full-connected graph for convenience reasons only.

Note that the best network from the second objective point of view will be always a very low connected (tree like) structure, which is indeed the cheapest option but also has the lowest resilience possible.

As the purpose of this study is to find a structure which shows high resilience to targeted attacks and low implementation cost, it is necessary to find an overall objective function which combines the two opposite objectives: robustness and cost. This issue will be discussed below, in the G.A. section.

## 3 Genetic Algorithm

As it was said above, the whole optimizing process is based on simple observation of network's structure. The genetic algorithm modifies graph's structure trying to find a particular network which provides the best resilience and the lowest implementation cost. The whole optimizing process is described below.

Genetic Algorithms usually uses populations of *individuals*. Each of the individuals has its own *performance level* which depends of how good it is as a potential solution of the given problem. The most effective individuals are allowed to reproduce themselves, usually through the most common genetic operators, such as crossover, mutation, cloning. Thus yield new individuals called *children* who keep inside them issues from their parents. The less effective individuals will die, while the effective ones will forward their capabilities to the next generation.

Genetic Algorithms have basically several specific elements, as follows: parameters encoding for the given problem, solution search limits, objective function used to select the best individuals for the reproduction and the hazard involved in the evolution.

### 3.1 Parameters Encoding

G.A. starts with an initial population made by *n* individuals, each individual being a Scale-Free network encoded using its adjacency matrix. Unlike the common genetic algorithms which use a binary array for parameters encoding, in this paper a 2D encoding is proposed. Representing the networks using their adjacency matrix will allow the genetic operators to work directly on the network's structure thus providing fast processing and convergence.

$$A_{ij} = \begin{cases} 1 & edge \\ 0 & no\_edge \end{cases} \tag{4}$$

The initial Scale-Free networks have been generated using *preferential attachment* method (Barabasi&Albert)[2,3,4].

Thus, the ten Scale-Free networks obtained are tested, the initial population is genetically modified and, trough the selection process, new generations yield, better from performance point of view.

### 3.2 Genetic Operators

The presented algorithm only uses mutation as genetic operator. Instead of using crossover and mutation as most of the Genetic Algorithms, two types of mutation have been chosen, each of them having its specific relevance for the network's evolution.

**Mutation1** provides a smooth movement in the space of solutions by making fine adjustments for finding a local optimum point. In order to do this, the operator randomly chooses one node *i*, take one of its stubs (element of the adjacency matrix) at random and modifies its value, or not, with equal probability.

**Mutation2** provides a wide search for the global optimum point by generating large jumps in the space of solutions. In order to do this, the operator randomly chooses one node *i*, take all of its stubs (each element of the adjacency matrix situated on the row/column *i*) and modifies their values, or not, with equal probability.

The new individuals obtained through the mutation process are reinserted in the old population using ROULETTE method [1] and based on their performance regarding the objective function. The objective function will be described below in paragraph 3.3.

### 3.3 Objective Function

As the problem is a multi-objective one, the objective function must be carefully chosen, so the evolutionary process could go into the right direction, "increase the robustness with low cost" that is. For a good fit of the objective function with this goal, the method of "weighting coefficients" has been chosen. This means, if there is a two variables function (this means two objectives), the two objectives can be virtually separated, by giving each of them its specific importance (weight) in the optimization process, like in (5) in general:

$$f(x_1, x_2) = w_1 \cdot f(x_1) + w_2 \cdot f(x_2) \tag{5}$$

or, like in (6) for this application:

$$f_{obj} = w \cdot f_{obj_1} + (1-w) \cdot s \cdot f_{obj_2} \tag{6}$$

If decided to minimize or maximize the overall objective function then both of the terms should be minimized or maximized, respectively.

As discussed above, in paragraph 2, there are two opposite goals to be accomplished in this application. First is to maximize the aria (2) and second, to minimize the normalized cost (3). This means that in the overall objective function one term is to be maximized and one to be minimized. This can be easily fixed by using the opposite value of one of the terms, as follows:

$$f_{obj} = w \cdot \frac{1}{f_{obj_1}} + (1-w) \cdot s \cdot f_{obj_2} \tag{7}$$

Now both of the terms have to be minimized, so the overall objective function will be also minimized. Minimizing the overall objective function $f_{obj}$ is the best option for this type of application as it provides in the end a very easy way to understand the results.

In (6,7) '*s*' is a scaling coefficient introduced there for bringing the two terms in the same variation range. Note that without coefficient '*s*' the second term would be always much smaller than the first one during the evolutionary process, and thus irrelevant for the optimization.

$$\frac{1}{f_{obj_1}} \cong s \cdot f_{obj_2} \tag{8}$$

As about the weighting coefficients, unlike the general use of them, only one single coefficient has been used, as 1's complement, in order to let one decide the importance of the two objectives for his specific application ($w \in [0,1]$).

### 3.4 Scenario and Process Flow

Number of networks per population is 10. Each network has Power-Law distribution of degrees, generated using Preferential Attachment. Each network has 500 nodes. The algorithms runs until 50 generations are processed. The genetic algorithm flows as follows:

*Compute initial population* $Pop_{init}$;

WHILE (number of generations < 50)

- *select individuals for reproduction;*
- *create offsprings using* mutation1*;*
- *create offsprings using* mutation2*;*
- *compute new generation (*reinsertion*);*

END

## 4 Results

In order to test the strength of the proposed GA, *w* was set to 1 and 0, thus disabling the second and the first objective respectively.

As it was expected, for *w*=1 only the first objective has been taken into account (Eq. 9), the GA evolving in the direction of improving the resilience of the network to

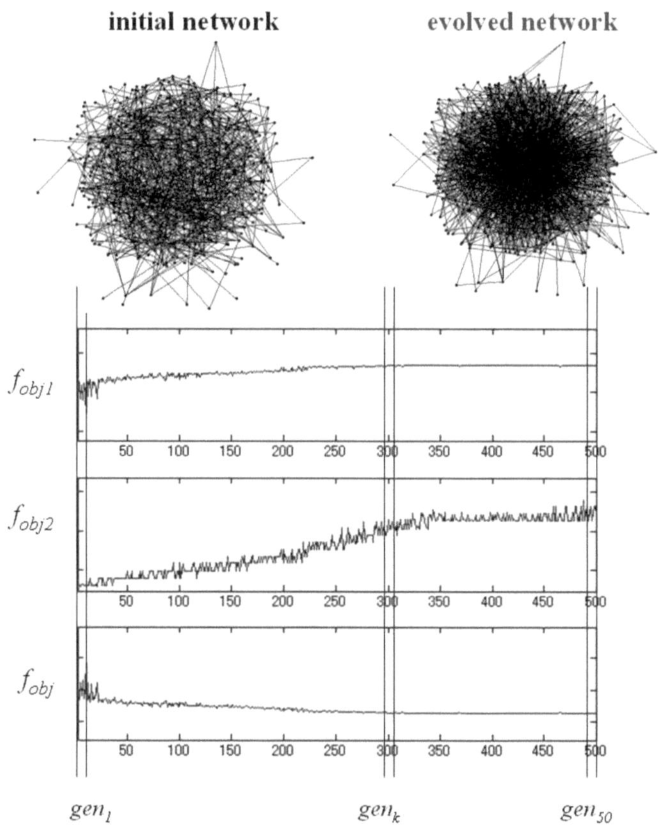


**Fig. 2.** TOP: Evolved network is very high connected. BOTTOM: first objective (robustness) increases. Second objective is not controlled by the GA and increases dramatically. The overall objective has the first objective's opposite variation.

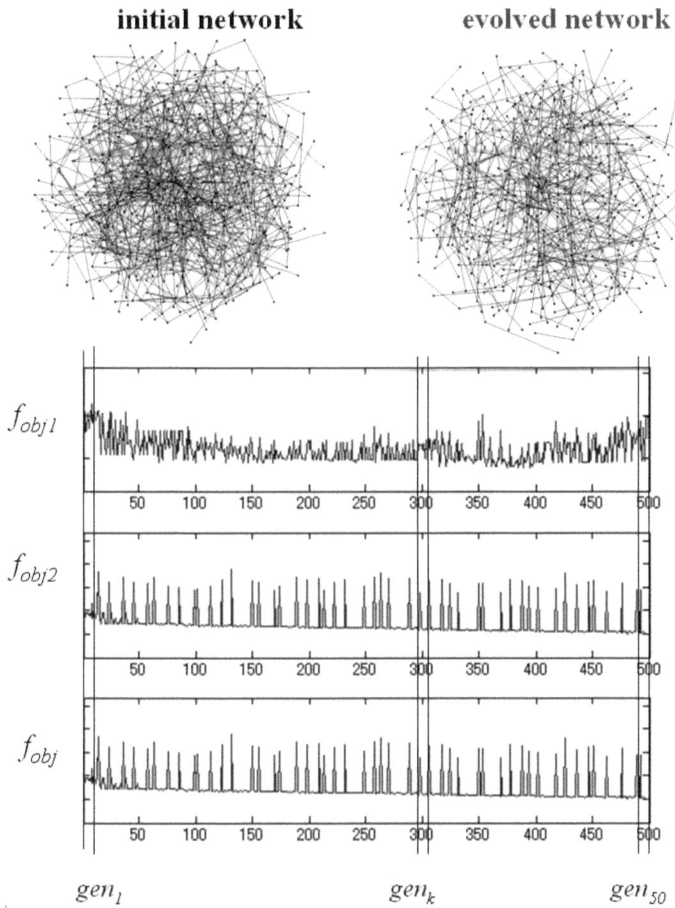


**Fig. 3.** TOP: Evolved network is very low connected (tree like). BOTTOM: first objective decreases, as it is not controlled by GA. Second objective is minimized. The overall objective is proportional with the second objective.

targeted attacks. As a result, the optimized network is very robust but also very high connected, as there was no restriction for the number of edges of the graph (Fig. 2).

$$f_{obj} = \frac{1}{f_{obj_1}} \tag{9}$$

For $w$=0 only the second objective has been taken into account (Eq. 10), the GA evolving in the direction of reducing the cost, which is in fact the connectivity of the graph. As a result, the optimized network is very cheap but also has low resilience, as there was no restriction for the robustness level (Fig. 3).

$$f_{obj} = s \cdot f_{obj_2} \tag{10}$$

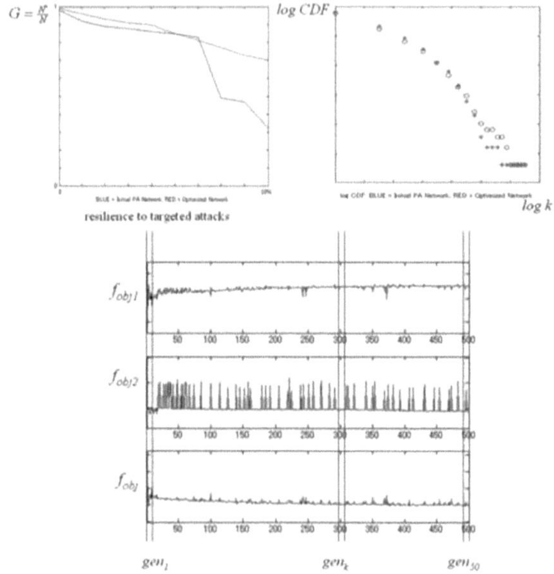


**Fig. 4.** TOP: circle/blue – initial network; star/red – evolved network. BOTTOM: evolution of the objective functions.

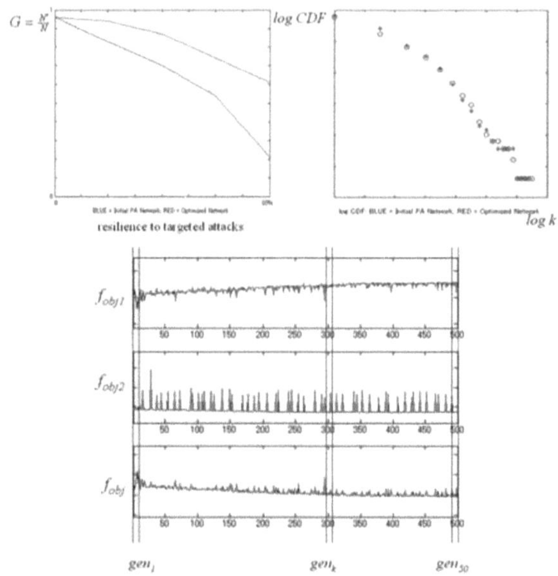


**Fig. 5.** TOP: circle/blue – initial network; start/red – evolved network. BOTTOM: evolution of the objective functions.

At this point the genetic algorithm has been tested and worked properly for each of the objectives. Still, the purpose of this application was to optimize graph's structure taking in account both constrains. For doing this, $w$ should be chosen between 0 and 1 depending on one's interest in either robustness or cost. Assuming that after the optimizing process the evolved network has to have the same type of structure as the initial one, *w* has been chosen between 0.3 and 0.5. After several simulations has been demonstrated that by choosing *w* in this interval, after 50 generation the network evolves in a more robust network, less connected than the initial one and having virtually the same type of degree distribution: Power-Law.

For *w*=0.5 both objectives have the same importance. As a result, the evolved network has a better resilience while the connectivity is forced to remain almost constant. The overall objective function decreases, showing that the GA evolves in the right direction (Fig. 4 Bottom). In the same time, the cumulative distribution function CDF of the evolved network is almost the same with the initial network's one (Fig. 4 Top).

For *w*=0.3 both objectives are taken into account, but the cost becomes more important than the previous case (*w*=0.5). As a result, the evolved network has a better resilience while the connectivity is forced to decrease. The overall objective function decreases, showing that the GA evolves in the right direction (Fig. 5 Bottom). In the same time, the cumulative distribution function CDF of the evolved network is again almost the same with the initial network's one (Fig. 5 Top).

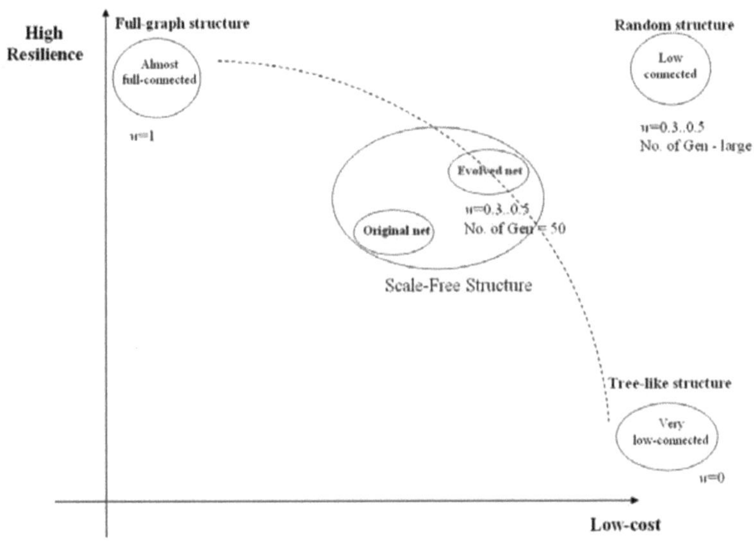


**Fig. 6.** For specific values of *w* and appropriate number of generations the original distribution of degrees can be preserved in some limits

## 5 Conclusions

First of all, this paper demonstrated that the proposed GA, with its new type of encoding system and genetic operators, could be one viable solution for designing or optimizing large scale networks. Taking into account both robustness and cost, strong networks can be found, depending on one's interest for either strength or cost.

Though, beyond this simple optimization issue there is more that can be said. It is well known and obvious that if the GA keeps running for a very large number of generations, the evolved network will be a random graph with low connectivity, in other words immune to targeted attacks and very cheap (Fig. 6). This is not such a good situation for the one who has a vulnerable Scale-Free network, and wants a resilient Scale-Free network. In the end he/she will need a more resilient network at the same type with the initial one. As an example, if had an internet type network and wanted to make it more robust, then turning it into a transportation type network wouldn't be such a good solution. For this reason, the most important achievement is that, that for specific values of $w$ and appropriate number of generations the original structure of the network can be preserved in some limits. Of course, structure means in this case degree distribution *only* and this starts a new question. If the evolved network has lower connectivity, higher resilience and virtually the same degree distribution, then other parameters must have been changed during the evolutionary process. If the nodes still have almost the same distribution of degrees then the way in which they are connected to each other has changed. For analyzing this issue, parameters like assortativity, modularity, cluster coefficient or betweenness must be taken in account for the future work.

## References


1. Goldberg, D.E.: Genetic algorithms in search, optimization and machine learning. Addison-Wesley, Reading (1989)
2. Barabasi, A.L., Albert, R.: Emergence of Scaling in Random Networks. Science 286, 509 (1999)
3. Barabasi, A.L., Albert, R., Jeong, H.: Mean-field theory for scale-free random networks. Physica A 272, 173–187 (1999)
4. Barabasi, A.L., Albert, R., Jeong, H.: Scale-free characteristics of random networks: the topology of the world-wide web. Physica A 281(1-4), 69–77 (2000)
5. Li, L., Alderson, D., Willinger, W., Doyle, J.: A First-Principles Approach to Understanding the Internet's Router-level Topology. IEEE Transactions on Networking, 1205–1218 (2005)
6. Boyan, J.A., Littman, M.L.: Packet Routing in dynamically Changing Networks: A Reinforcement Learning Approach. Advances in Neural Information Processing Systems 6 (1994)
7. Newth, D., Ash, J.: Evolving cascading failure resilience in complex networks. In: Proceedings of The 8th Asia Pacific Symposium on Intelligent and Evolutionary Systems, Cairns, Australia (2004)
8. Motter, A.E.: Cascade control and defense in complex networks. Physics Rev. Lett. 93, 098701 (2004)
9. Motter, A.E., Lai, Y.C.: Cascade-based attacks on complex networks. Physics Rev. E 66, 065102 (2002)
10. Lai, Y.C., Motter, A.E., Nishikawa, T.: Attacks and Cascades in Complex Networks. Physics Lectures Notes, vol. 650, p. 299 (2004)
11. Leu, G., Namatame, A.: Efficient recovery from cascade failures. In: Proc. of the 10th Asia Pacific Workshop on Intelligent and Evolutionary Systems, South-Korea (2006)

12. Barrat, A., Barthelemy, M., Pastor-Satorras, R., Vespignani, A.: Proc. Natl. Acad. Sci., USA 101, 3747 (2004)
13. Barrat, A., Barthelemy, M., Vespignani, A.: J. Stat. Mech., P05003 (2005)
14. Dall'Asta, L., Barrat, A., Barthelemy, M., Vespignani, A.: Vulnerability of weighted networks. DELIS (2006)
15. Dorogovtsev, S.N., Mendes, J.F.F.: Evolution of Networks: from biological nets to the Internet and WWW. Oxford University Press, Oxford (2003)
16. Cohen, R., Erez, K., ben-Avraham, D., Havlin, S.: Physics Lett. 85, 4626 (2000)
17. Callaway, D.S., Newman, M.E.J., Strogatz, S.H., Watts, D.J.: Physics, Lett. 85, 5468 (2000)
18. Holme, P., Kim, B.J., Yoon, C.N., Han, S.K.: Physics, E 65, 056109 (2002)
19. Newman, M.E.J.: The mathematics of networks, 2nd edn. The New Palgrave Encyclopedia of Economics

# Evolving Networks with Enhanced Linear Stability Properties

David Newth[1] and Jeff Ash[2]

[1] CSIRO Centre for Complex Systems Science,
CSIRO Marine and Atmospheric Research
david.newth@csiro.au

[2] Centre for Research into Complex Systems (CRiCS), Charles Sturt University
jash@csu.edu.au

Networks are so much a part of our modern society that when they fail the effects can be significant. In many cases, global network failures can be triggered by seemingly minor local events. Increased understanding of why this occurs and, importantly, the properties of the network that allow it to occur, is thus desirable. In this account we use an evolutionary algorithm to evolve complex networks that have enhanced linear stability properties. We then analyze these networks for topological regularities that explain the source their stability/instability. Analysis of the structure of networks with enhanced stability properties reveals that these networks have a highly skewed degree distribution, very short path-length between nodes, have little or no clustering and are dissasortative. By contrast, networks with enhanced instability properties have a peaked degree distribution with a small variance, have long path-lengths between nodes, contain a high degree of clustering and are highly assortative. We then test the topological stability of these networks and discover that networks with enhanced stability properties are highly robust to the random removal of nodes, but highly fragile to targeted attacks. Networks with enhanced instability properties are robust to targeted attacks. These network features have implications for the physical and biological networks that surround us.

## 1 Introduction

Our modern society has come to depend on large-scale infrastructure networks to deliver resources to our homes and businesses in an efficient manner. Over the past decade, there have been numerous examples where a local disturbance has lead to the global failure of critical infrastructure. For instance, on August 10, 1996 in Oregon a combination of hot weather and abnormally high electricity demand caused power lines to sag into trees and trigger a cascade failure of power stations, distribution substations, and assorted other infrastructure which affected power supplies to 11 states [1]. On August 14, 2003 a similar train of events starting in Ohio triggered the largest blackout in North American history [2]. Australia and New Zealand have not been left untouched. In Auckland

M. Gen et al.: Intelligent and Evolutionary Systems, SCI 187, pp. 61–77.
springerlink.com 

the failure of four major distribution cables began on January 22, 1998, and when the last of these collapsed almost a month later on February 20 the city was left totally without power. Seventeen days later this city had still only managed to regain 40% of its capacity [3]. Where a network is carrying a flow of some particular resource (electricity, gas, data packets, information, etc.) nodes individually experience a load, and under normal circumstances this load does not exceed the capacity of that node. Nodes also have the ability to mediate the behavior of the network in response to a perturbation (such as the failure of a neighboring node or a sudden local increase in flow). Critical infrastructure are continually confronted with small perturbations. Most of these disturbances have no effect on the network's performance overall. However, a small fraction of these disturbances cascade through the network, crippling its performance. The resilience of a network to the propagation of disturbances is directly related to the underlying topology of the network.

In previous work [4] we have examined the topological properties of networks that make them resilient to cascading failures. In this account we use a search algorithm to help us identify network properties that lead to enhanced linear stability properties. We also show that networks that display enhanced linear stability properties are highly resilient to the random loss of nodes. By contrast networks with enhanced instability properties tend to be more resilient to the loss of specific (or important) nodes.

The remainder of this paper is organized as follows. In the next section, we define the network properties that are used to evaluate the stability of a given network. In Section 3 we describe the rewiring algorithm and the experimental setup used. Section 4 outlines the network properties of interest to us here. In Sections 5 and 6 we provide high level descriptions of networks having enhanced stability, and enhanced instability properties, and how these networks change over time. Section 7 provides a systematic analysis of the structural properties of the evolved networks. Section 8 examines the topological stability of the evolved networks. Finally in Section 9 we discuss the major findings and future directions of this work.

## 2 Stability Analysis of Complex Networks

Many of the complex systems that surround us — such as power grids, food webs, social systems, critical infrastructure, traffic flow systems, the internet and even the brain — are large, complex, and grossly non-linear in their dynamics. Typically, models of these systems are inspired by equations similar to:

$$\frac{dX_i}{dt} = F_i\left(X_1(t), X_2(t), \ldots, X_n(t)\right), \tag{1}$$

where $F_i$ is an empirically inspired, nonlinear function of the effect of the $i^{th}$ system element on the dynamics of the other $n$ system elements. When modeling ecological systems, the function $F_i$ takes on the form of the Lotka-Volterra equations [5, 6]:

$$F_i = X_i \left( b_i - \sum_{j=1}^{n} X_j \alpha_{ij} \right), \tag{2}$$

where $X_i$ is the biomass of the $i^{th}$ species; $b_i$ is the rate of change of the biomass of species $X_i$ in the absence of prey and predators; and $\alpha_{ij}$ is the pre unit effect of species $j$'s biomass on the growth rate of species $i$'s biomass. In other applications, $F_i$ can take on the form of non-linear oscillators [7]; the Black-Scholes equations [8]; or non-linear chemical reactions [9]. Of particular interest is the steady state of the system, in which all growth rates are zero, giving the fixed point or steady state values for each of the control variables $X_i^*$. This occurs when:

$$0 = F_i \left( X_1^*(t), X_2^*(t), \ldots, X_n^*(t) \right). \tag{3}$$

The local dynamics and stability in the neighborhood of the fixed point can be determined by expanding equation 1 in a Taylor series about the steady state:

$$\frac{dX_i(t)}{dt} = F_i|_* + \sum_{j=1}^{n} \left[ \frac{\partial F_i}{\partial X_j} \bigg|_* x_j(t) + \frac{1}{2} \sum_{k=1}^{n} \left[ \frac{\partial^2 F_i}{\partial X_j \partial X_k} \bigg|_* x_j(t) x_k(t) + \ldots \right.\right. \tag{4}$$

where $x_i(t) = X_i(t) - X_i^*$ and $*$ denotes the steady state. Since $F_i|_* = 0$, and close to the steady state the $x_i$ values are small, all terms that are second order and higher need not be considered in determining the stability of the system. This gives a linearized approximation that can be expressed in matrix form as:

$$\frac{d\mathbf{x}(t)}{dt} = A\mathbf{x}(t), \tag{5}$$

where $\mathbf{x}(t)$ is an $n \times 1$ column vector of the deviations from the steady state and the matrix $A$ has elements $a_{ij}$:

$$a_{ij} = \frac{\partial F_i}{\partial X_j} \bigg|_*, \tag{6}$$

which represents the effect of the variable $X_j$ on the rate of change of variable $i$ near the steady state. As May demonstrates [10], solving the following equation for $A$ reveals the temporal behavior of the system.

$$(A - \lambda I)\mathbf{x}(t) = 0. \tag{7}$$

Here $I$ is the $m \times m$ unit matrix. This set of equations possesses a non-trivial solution if and only if the determinant vanishes:

$$\det|A - \lambda I| = 0. \tag{8}$$

This is in effect the $m^{\mathrm{th}}$ order polynomial equation in $\lambda$, and it determines the eigenvalues $\lambda$ of the matrix $A$. In general they are complex numbers

$\lambda = \zeta + i\xi$, with the real part $\zeta$ producing exponential growth or decay, and $\xi$ producing sinusoidal oscillations. In this account we are only interested in the real part of the eigenvalues. The eigenvalues of a system can be ordered: $|\lambda_1| < |\lambda_2| < \ldots < |\lambda_{n-1}| < |\lambda_{\max}|$, and we will refer to $\lambda_{\max}$ as the dominant eigenvalue. If the real part of $\mathrm{Re}(\lambda_{\max}) < 0$, then the system is said to be stable to perturbations in the region of the fixed point. Here, we will evolve networks that have enhanced stability properties (i.e $\min(\mathrm{Re}(\lambda_{\max}))$), and networks that have enhanced instability properties (i.e $\max(\mathrm{Re}(\lambda_{\max}))$).

## 3 Evolving Complex Networks

Now we develop a search algorithm to find adjacency matrices $A$ with enhanced stability properties. To do this we make use of a stochastic hill climber. The re-wiring scheme adopted here is similar to others used in previous studies [4, 11], and the effectiveness of this algorithm at finding networks with enhanced stability properties is given in [11].

The optimization scheme consists of three steps: (1) an edge is selected and one end of the edge is reassigned to another node; (2) the dominant eigenvalue ($\lambda'_{\max}$) is calculated for the modified network; and (3) if $\lambda'_{\max}$ is superior to $\lambda_{\max}$ of the original network, the rewiring is accepted, else the edge re-wiring is rejected. These three steps are repeated for $10^5$ time steps. The eigenvalues were determined numerically with routines from the numerical recipes in C [12]. The networks studied here consist of 100 nodes and 150 edges, and initially the algorithm was seeded with an Erdös-Rényi random graph [13]. The edges were set to a value of 1, but this can be easily modified to take on real values. By convention, the on-diagonal or self-regulating terms were set to -1. At every step, the network was checked to ensure it consisted of a single connected component.

## 4 Structure of Complex Networks

The matrix $A$ defines an adjacency matrix that describes the interactions between elements within the system. The patterns of interactions between system elements form a complex network. Over the past 10 years, complex networks from widely varying domains have been shown to share common statistical properties. These properties include short path length and high clustering (the so-called small-world properties), assortativity and scale-free degree distribution. The remainder of this section describes each of these properties.

### 4.1 Small-World properties

Small-world properties [14] can be detected through two statistics, the average shortest-path length and the clustering coefficient. The average shortest-path length ($l$) is defined as:

$$l = \frac{1}{N(N-1)} \sum_{i=1}^{N} \sum_{j=i+1}^{N} l_{\min}(i,j), \tag{9}$$

where $l_{\min}(i,j)$ is the shortest path distance between nodes $i$ and $j$, and $N$ is the number of nodes. The diameter of a network ($l_{\max}$) is the longest shortest-path between two nodes within a network.

Clustering is a common feature found in many networks. The degree of clustering within a network is captured by the clustering coefficient. Given a node $N_i$ with $k_i$ neighbors, $E_i$ is defined to be number of links between the $k_i$ neighbors. The clustering coefficient is the ratio between the number of links that exists between the neighbors of $N_i$ and the potential number of links $k_i(k_i-1)$. The average clustering coefficient is:

$$C = \frac{1}{N} \sum_{i=1}^{N} \frac{2E_i}{k_i(k_i-1)}. \tag{10}$$

A network is said to have small-world properties if, compared to an Erdös-Rényi random graph, the following conditions hold: $l \approx l_{\text{rand}}$ and $C \gg C_{\text{rand}}$ [14].

### 4.2 Assortativity

A network displays assortative mixing if the nodes in the networks that have many connections tend to be connected to other nodes having many connections. A network is said to be disassortative if the highly connected nodes tend to be connected to nodes having few connections. The degree of assortativeness can be detected through the use of the Pearsons correlation coefficient. Such correlations can be defined as

$$\alpha = \frac{c\sum_i j_i k_i - \left[c\sum_i \frac{1}{2}(j_i+k_j)\right]^2}{c\sum_i \frac{1}{2}(j_i^2+k_i^2) - \left[c\sum_i \frac{1}{2}(j_i+k_j)\right]^2}, \tag{11}$$

where $j_i$ and $k_i$ are the degrees of the vertices at the ends of the $i^{th}$ edge. The constant $c$ is defined as the reciprocal of $m$ where $m$ is the number of edges i.e. $c = \frac{1}{m}$. A network displays assortative mixing when $\alpha < 0$ and disassortative mixing when $\alpha > 0$. Studies have found that social networks display assortative mixing, while systems with a power-law degree distribution are disassortatively mixed [15].

### 4.3 Degree Distribution

One of the common structural properties found in many man-made and natural complex networks is a degree distribution with a power-law tail $P(k) \approx k^{-\gamma}$, where exponent $\gamma$ is in the range between 2 and 3 [16]. The degree of a node is the number of links possessed by that node. Networks exhibiting these power-law degree distributions are known as scale-free networks. Several mechanisms have been proposed for the formation of such topological features. Albert and Barabási [16] showed that a preferential attachment mechanism leads to a degree distribution with a power-law tail. Ferrer-Cancho and Solé [17] showed that the minimization of path length and the number of links contained within a network also leads to scale-free structures. These results suggest that scale-free networks may be an efficient and stable configuration for many complex networks.

# 5 Evolving Stable Networks

In the first experiment, we evolved networks with enhanced stability properties, that is we attempted to minimize the dominant eigenvalue. This experiment was repeated 200 times, and the resulting network from each run was collected

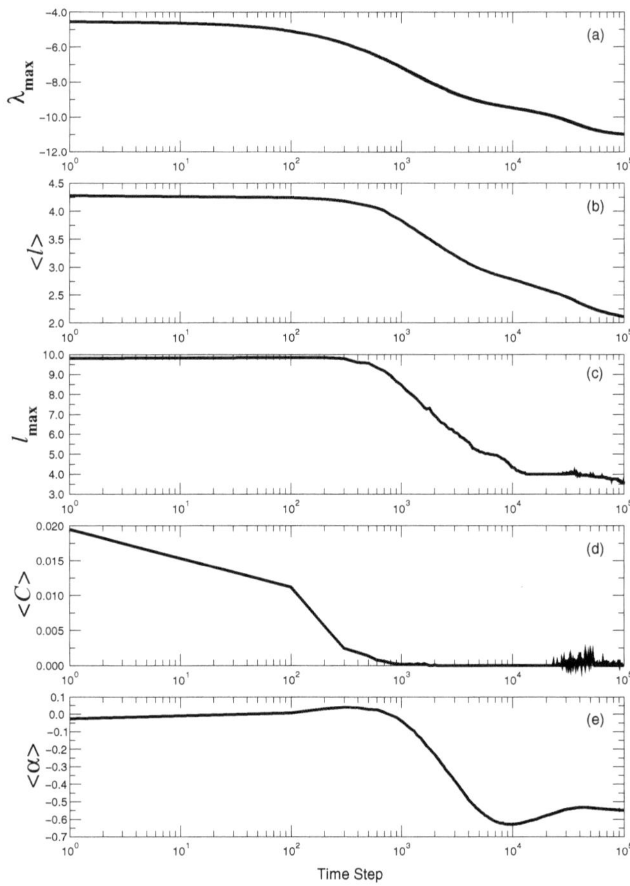


**Fig. 1.** Time evolution of networks with enhanced stability properties. From top to bottom: Time evolution of $\lambda_{\max}$; time evolution of the average shortest-path length ($\langle l \rangle$); time evolution of the network diameter ($l_{\max}$); time evolution of the clustering coefficient ($\langle C \rangle$); and finally time evolution of assortative mixing ($\langle \alpha \rangle$).

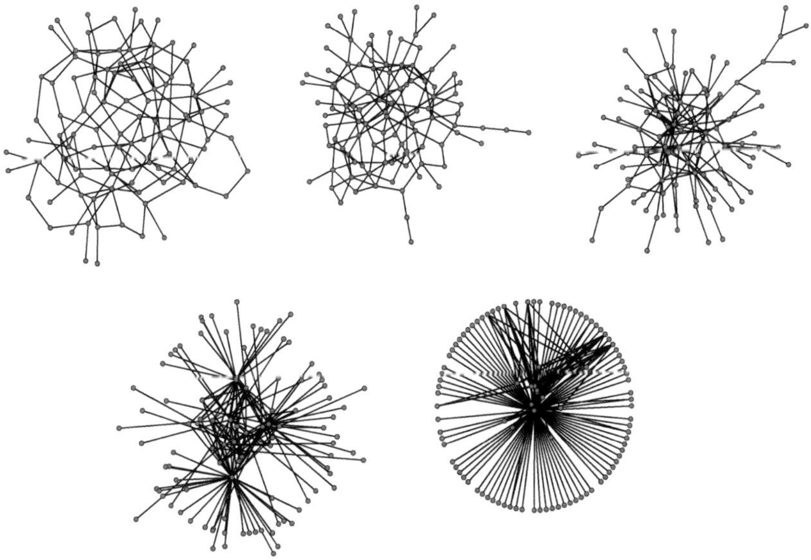

**Fig. 2.** Structural changes as the networks evolve increased stability properties. From top left to bottom right, example networks at time $t = 0$, $t = 100$, $t = 1000$, $t = 10000$ and $t = 100000$. As the system evolves, the most striking feature is the emergence of hub-like structures that form star-like networks. These structures account for the structural properties outlined in Fig. 1.

and analyzed in more detail (see Section 7). Figure 1 shows the time evolution of the complex networks. From Figure 1 it can be seen that as the networks become more stable, the average shortest-path length, diameter, degree of cluster and assortativity all decrease. In searching for more stable configurations, it appears that short cycles (ie clustering) are removed first. As a visual illustration, Figure 2 shows example networks after 0, 100, 1000, 10000 and 100000 rewiring steps.

## 6 Evolving Unstable Networks

In the second experiment, we evolved networks with enhanced instability properties, that is we attempted to maximize the dominant eigenvalue. This experiment was repeated 200 times, and the resulting network from each run was collected and analyzed in more detail (see Section 7). Figure 3, shows the time evolution of the complex networks. It can be seen that as the networks become more unstable, the average shortest-path length, diameter, degree of cluster and assortativity all increase. In searching for more unstable configurations, it appears that short and long cycles become dominant features of these networks. As a visual illustration, Figure 4 shows example networks after 0, 100, 1000, 10000 and 100000 rewiring steps.

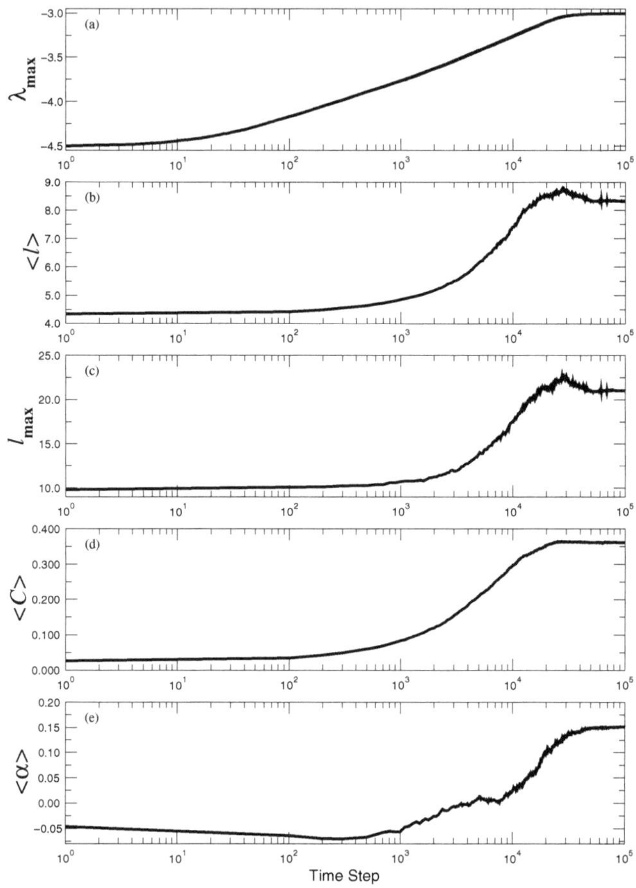


**Fig. 3.** Time evolution of networks with enhanced instability properties. From top to bottom: Time evolution of $\lambda_{max}$; time evolution of the average shortest-path length ($\langle l \rangle$); time evolution of the network diameter ($l_{max}$); time evolution of the clustering coefficient ($\langle C \rangle$); and finally time evolution of assortative mixing ($\langle \alpha \rangle$).

## 7 Topological Properties of Evolved Networks

Extensive studies of the degree distribution of real-world networks have identified three main classes of networks: (1) scale-free networks, characterized by a vertex connectivity distribution that decays as a power law; (2) broad-scale networks, characterized by a connectivity distribution that has a power-law regime followed by a sharp cut-off; and (3) single-scale networks, characterized by a connectivity distribution with a fast decaying tail [18]. Figure 5 shows the degree distribution for the networks evolved with enhanced stability and enhanced instability properties. The degree distribution for networks with enhanced stability properties is heavily skewed when compared to the initial random networks (dashed lines).

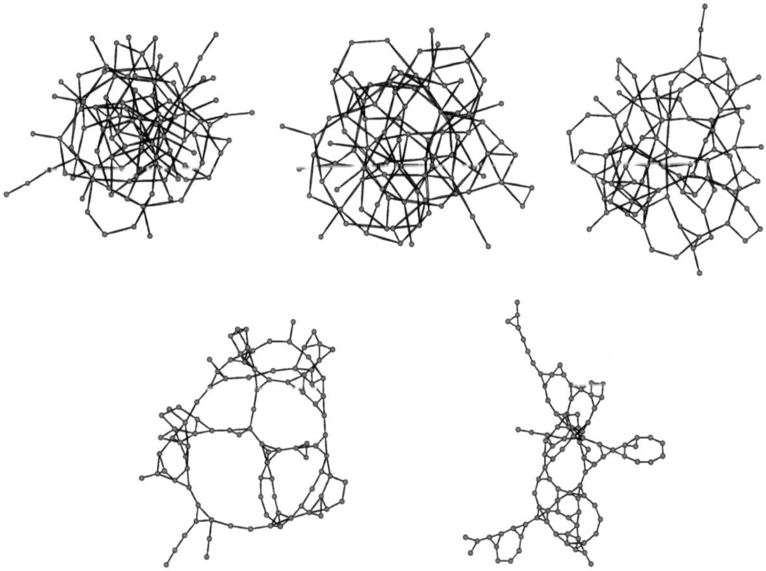

**Fig. 4.** Structural changes as the networks evolve increased instability properties. From top left to bottom right, example networks at time $t = 0$, $t = 100$, $t = 1000$, $t = 10000$ and $t = 100000$. As the system evolve, the most striking feature here is the formation of long loop and path structures, (see Fig. 3).

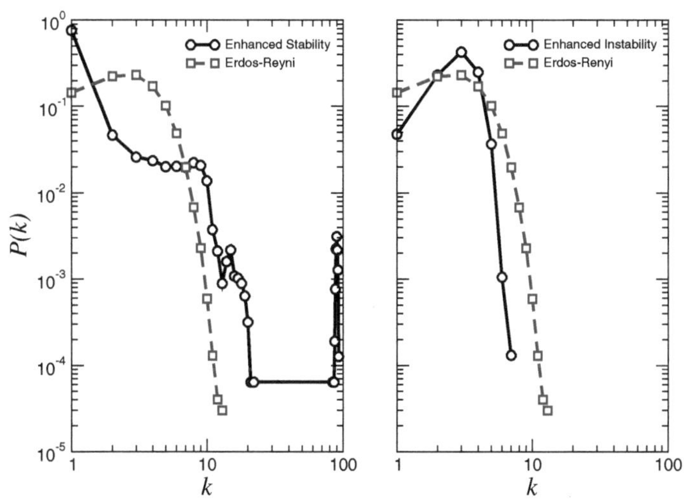


**Fig. 5.** Degree Distributions. (Left) Degree distribution for networks with enhanced stability properties; (Right) Degree distributions for networks with enhanced instability properties.

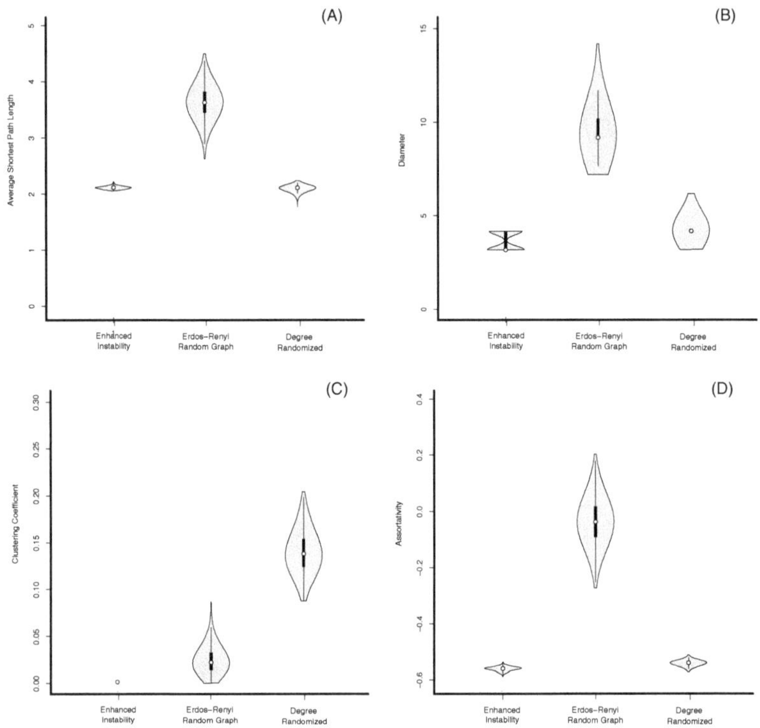


**Fig. 6.** Comparison between the evolved networks with enhanced stability properties and the two random null models. (A) Average shortest-path length; (B) Diameter; (C) Clustering Coefficient; and (D) Assortativeness.

Despite the short tail (which is due to finite size effects), there is a significant fraction of nodes with large degrees. This indicates that the resulting network is quite inhomogeneous. The degree distribution for networks with enhanced instability properties by contrast is quite peaked, with a narrower variance than the random initial conditions. This suggests that networks with enhanced instability have a degree of regularity about the way links are distributed through the networks.

In previous studies [16, 19, 20], it has been highlighted that many of the statistical properties of a network are derived directly from the degree distribution. In an attempt to determine how unique or special these evolved networks are we have compared their network statistics to those of two random null models. The first of these is an Erdös-Rényi random graph to determine those characteristics which can be accounted for purely by random interactions. The second null model is the degree randomization model as described in [20]. This model assumes that the degree distribution is the source of the network properties. The model randomizes node connection (i.e. which node is connected to which other node), but preserves the individual node degree characteristics. Comparison

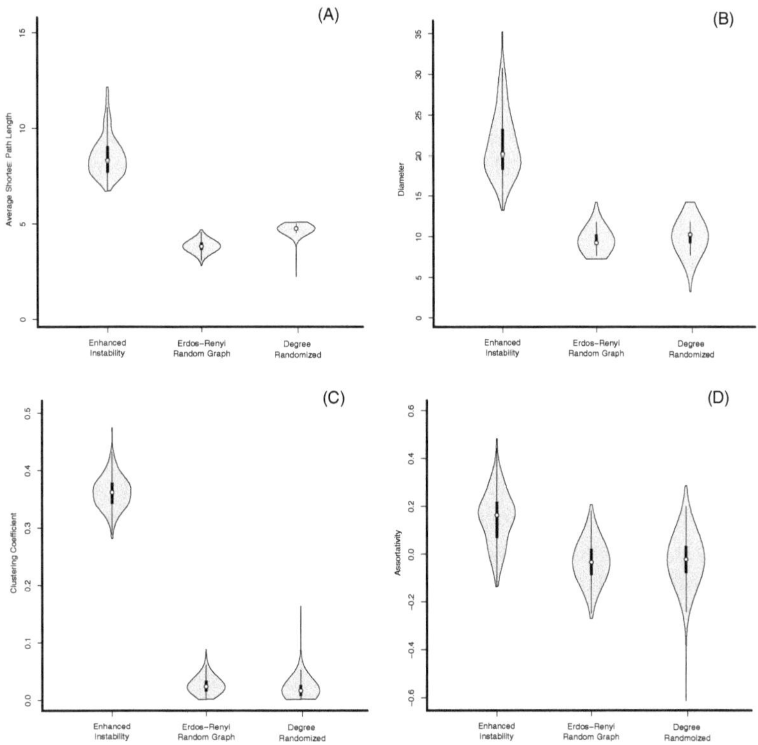


**Fig. 7.** Comparison between the evolved networks with enhanced instability properties and the two random null models. (A) Average shortest-path length; (B) Diameter; (C) Clustering Coefficient; and (D) Assortativeness.

between the evolved networks and the two null models, shows what is unique to the evolved networks, as well as what properties can be accounted for by random assemblage and the degree distribution.

Figure 6 shows the comparison between the evolved networks with enhanced stability properties and the two null models. The plots show summary statistics for each of the network characteristics of interest. The plots used here are violin plots [21], which include all the information found in a box plot, but also include the density trace. This provides more information about the structure of the underlying distribution. In all cases, the violin plots were drawn from statistics taken from 1000 null models. Figures 6(A) and 6(B) show the variation in the average shortest-path length and diameter respectively. In both cases these characteristics are not significantly different from the network characteristics of the degree randomized network. This indicates that these characteristics are directly related to degree distribution. Figure 6(C) compares the clustering across the observed and null models. The evolved networks have no clustering, unlike the two null models. The lack of clustering is a unique characteristic of the evolved networks. Finally figure 6(D) shows the assortativity of the evolved networks and the null models. The evolved networks are highly disassortative. The disassortivity of the evolved networks is similar to the level of disassortativity found in the degree randomized model. The assortativity observed in the evolved networks is a direct result of the degree distribution. In short the degree distribution accounts for the path-length characteristics and assortativity in the evolved networks. However, the degree of clustering is a unique property of the evolved networks.

Figure 7 shows a comparison between the evolved networks with enhanced instability properties and the two null models. Figure 7(A) and 7(B) shows the variation in the average shortest-path length and diameter respectively. In both cases these characteristics are significantly larger than those found in the two null models. This suggests that the path length characteristics are unique to the evolved class of networks. The clustering (Figure 7(C)) found in the evolved networks is significantly higher than that observed in the random null models. Combined, the high clustering and long average shortest-path length characteristics suggest that these networks have the so-called "long-world" characteristics. Finally networks with enhanced instability properties tend to be assortative (Figure 7(D)), however the spread of these distributions is wide. In the case of the networks with enhanced instability properties, it appears that the degree distribution does not account for the increased clustering, assortativity and path-length characteristics. The evolved networks have a somewhat "unique" wiring which gives them increased modularity, and clustering than that which is given solely by their degree distribution.

## 8 Resilience to Topological Attack

Now, we examine the topological resilience [19] of the evolved networks. By topological resilience, we mean how these networks break apart when nodes are removed from the network (attacked). The following section outlines the strategies used to select nodes that are to be removed from the networks. We then examine the response of the evolved networks to each of these attack regimes.

### 8.1 Node Removal Schemes

Here we consider four node removal schemes, to represent a number of attack scenarios experienced by real world networks in different situations. These four are: (1) Random node removal; (2) Degree centrality node removal; (3) Betweenness centrality node removal; and (4) Closeness centrality node removal. Under the random node removal scheme, nodes are removed from the network without bias. The other schemes target nodes based on node centrality. After a node is removed from the network, the centrality of each node is recalculated. If two nodes have the same centrality score, the node selected to be removed from the network is chosen at random. Each of the centrality measures used to remove nodes is outlined below.

- *Degree Centrality Node Removal.* The first targeted node removal scheme removes nodes based in their degree centrality. Degree centrality $C_D(v)$ of node $v$ is defined as the number of links incident upon a node (i.e., the number of edges or neighbors $(k_n)$ that node $v$ has):

$$C_D(v) = k_v. \tag{12}$$

Degree centrality is often interpreted in terms of the immediate "risk" to a node from whatever is flowing through the network (for example the chance of a node being "infected" in a situation where the network is modeling the spread of a virus);
- *Betweenness Centrality Node Removal.* The second targeted node removal scheme removes nodes based on their betweenness centrality. Betweenness centrality is a measure of a node's role in the transmission of information along shortest paths. The betweenness centrality $C_B(v)$ of vertex $v$ is:

$$C_B(v) = \sum_{\substack{s \neq v \neq t \in V \\ s \neq t}} \frac{\sigma_{st}(v)}{\sigma_{st}}, \tag{13}$$

where $\sigma_{st}$ is the number of shortest paths from $s$ to $t$ and $\sigma_{st}(v)$ is the number of shortest paths from $s$ to $t$ that pass through a vertex $v$. Betweeness centrality can be interpreted as a node's ability to mediate the flow of resources across the network (for example the effect on the movement of trains across a rail network, if a particular station (node) is experiencing heavy delays); and
- *Closeness Centrality Node Removal.* The final targeted removal scheme is based on node closeness centrality. Closeness centrality $C_C(v)$ of node $v$ is defined as the mean shortest-path distance between the node $v$, and all other vertices reachable from it:

$$C_C(v) = \frac{\sum_{\substack{t \in V \\ t \neq v}} l_{\min}(v,t)}{n-1}. \tag{14}$$

where $l_{\min}(i,j)$ is the shortest path distance between nodes $i$ and $j$. Closeness centrality can be thought of as a measure of how long it will take information to spread from a given vertex to other reachable vertices in the network (for example, how easy it is for data to travel from a source to a destination on a computer network).

## 8.2 Topological Stability

To test the topological stability of networks with enhanced properties we took 200 optimized networks from both schemes and subjected each network to the attack regimes outlined above in 8.1. This was repeated 100 times for each network, to allow for adequate selection between tied centrality measures. As nodes were removed we kept track of a number of statistics about the network. Each statistic was recorded as a function of the number of nodes removed from the

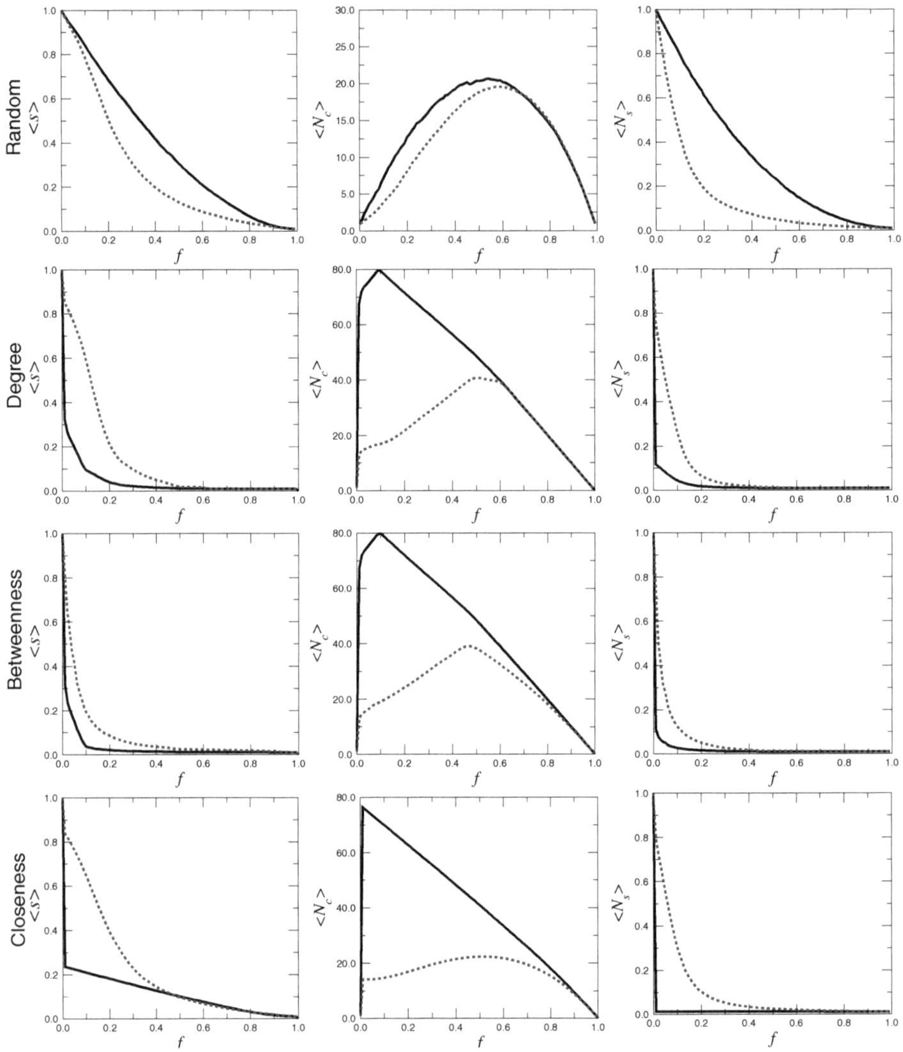


**Fig. 8.** Decay of networks as nodes are randomly and targetedly removed from the network. Solid lines are networks with enhanced stability properties, and dashed lines are networks with enhanced instability properties.

network $f$. The first statistic is the fraction of nodes in the largest connected component $\langle s \rangle$. The second statistic, $N_c$ is the number of clusters or connected components making up the network. Finally we kept track of the statistic $N_s$ which is the average cluster size. This is the number of nodes within the network, divided by the number of connected components $N_c$. Figure 8 illustrates how the two classes of network respond as nodes are removed.

From Figure 8 we can see how the networks break up as nodes are removed in accordance with each of the schemes. The most striking observation here is that networks with enhanced stability properties are highly fragile to targeted node removal — they quickly break apart into a number of very small clusters. But these networks are highly resilient to random node removal. By optimizing for dynamic stability, topological stability to random attack is gained at no cost. The networks with enhanced instability properties are less resilient to random node removal than networks with enhanced stability properties, but these networks are more resilient to targeted node removal. However, it still only requires the removal of several key nodes to break the networks into a number of disconnected components.

## 9 Discussion

In this paper, we have employed the use of an optimization algorithm to identify network characteristics that seem to be associated with enhanced linear stability and instability properties. Figures 2 and 4 show that the optimized networks display a degree of structural regularity in their arrangement. Networks with enhanced stability properties take on a star like structure. Hubs play an important role in many large scale infrastructure networks. While finite size effects make it difficult to determine the exact role and configuration of hubs that make these networks more stable, we postulate that the hubs allow perturbations to be distributed and reabsorbed quickly. However, a systematic test needs to be developed to gain a full understanding of the interconnected nature of the hubs. We can also make the following general observations about the networks with enhanced stability properties:

- Networks with enhanced stability properties have very low cluster, and almost no cycles;
- Networks with enhanced stability properties have a highly skewed degree distribution. The degree distribution accounts for many of the observed networks;
- Networks with enhanced stability properties tend to have short paths connecting any two nodes, a small diameter, and tend to be highly disassortative; and
- Networks with enhanced stability properties are highly resilient to random attack, but highly sensitive to targeted attack.

In addition, one of the interesting observations shown in Section 8 is that the networks with increased stability properties are also topologically stable — that is they tend not to fall apart — when nodes are randomly removed. However, these networks are vulnerable to targeted attacks. It is tempting to suggest that when a network is optimizing for stability, topological stability to random failure is obtained as a no-cost bonus.

From the work presented here we make the following observations about the networks with enhanced instability properties:

- Networks with enhanced instability properties have an interlocked loop structure;
- Networks with enhanced instability properties have a peaked degree distribution, and the degree distribution is not the sole source of the structural properties observed within these networks;
- Networks with enhanced instability properties tend to have longer average shortest-path lengths, larger diameter, higher clustering and tend to be more assortative than random null models; and
- Networks with enhanced instability properties are resilient to random and targeted attacks.

Many biological, social and large scale infrastructure networks display a surprising degree of similarity in their overall organization. Although these systems may look structurally similar, the origins of the similarity may be quite different. Biological networks, for example, exploit homeostasis provided by certain network properties, while technological networks arrive at the same properties as the result of a trade-off between communication efficiency and link cost [22]. For the simple system dynamics studied here, we suggest that modular design and clustering are key properties when designing complex networks that need to be robust to perturbations. But it should be noted that the mere observations of the characteristics outlined here does not imply increased stability or instability.

Finally, the work presented here opens a number of additional lines of study, and three deserve mention: (1) The networks studied here are all homogeneous. How does the system organize itself, where certain key components are more stable/unstable? (2) If capacity of a given node to regulate itself is measured in terms of a cost function, what configuration generates the most robust topology, while minimizing cost? (3) Many natural systems display a high degree of homeostasis. How do these networks compare with the evolved networks, and to large scale infrastructure networks? What are the sources (reasons) for the variations? All of these questions require further experimentation but can be explored in the context of the framework proposed here.

## References


1. CNN Interactive, sagging power lines, hot weather blamed for blackout, `http://www.cnn.com/US/9608/11/power.outage/`
2. US-Canada power system outage task force, Final Report on the August 14th blackout in the United States and Canada, `https://reports.energy.gov/BlackoutFinal-Web.pdfS`
3. Davis, P.: Earth Island J. 15(4) (2004)
4. Ash, J., Newth, D.: Phys A 380, 673–683 (2007)
5. Lotka, A.J.: Elements of physical biology. Williams and Wilkins Co., Baltimore (1925)
6. Volterra, V.: Mem. R. Accad. Naz. dei Lincei. 2 (1926)
7. Kuramoto, Y.: Chemical Oscillations, Waves and Turbulance. Springer, Berlin (1984)
8. Black, F., Scholes, M.: J. Polit Econ. 81(3), 637–654 (1973)

9. Kondepuid, D., Prigogine, I.: Modern Thermodynamics. Wiley, Germany (2002)
10. May, R.M.: Stability and Complexity in Model Ecosystems. Princeton University Press, Princeton (2001)
11. Newth, D., Brede, M.: Compl. Sys. 16(4), 100–115 (2006)
12. Press, W.H., Teukolsky, S.A., Vetterling, W.T., Flannery, B.P.: Numerical recipes in C. The art of scientific computing. Cambridge University Press, Cambridge (1992)
13. Erdös, P., Rényi, A.: Publ. Math. 6, 290–297 (1959)
14. Watts, D., Strogatz, S.: Nature 393, 440–442 (1998)
15. Newman, M.E.J.: Eur. Phys. J. B 38, 321–330 (2004)
16. Albert, R., Barabási, A.L.: Rev. Mod. Phys. 74, 247–297 (2002)
17. Ferrer-Cancho, R., Solé, R.V.: Optimization in complex networks. In: Statistical Mechanics of Complex Networks. Lecture Notes in Physics, pp. 114–125 (2003)
18. Amaral, L.A.N., Scala, A., Barthe lemy, M., Stanley, H.E.: Classes of small-world networks. Proc. Natl. Acad. Sci. 97(21), 11149–11152 (2000)
19. Albert, R., Jeong, H., Barabási, A.L.: Nature 406, 378 (2000)
20. Milo, R., Shen-Orr, S., Itzkovitz, S., Kashtan, N., Chklovskii, D., Alon, U.: Science 298, 824–827 (2002)
21. Hintze, J.L., Nelson, R.D.: The American Statistician 52(2), 181–184 (1998)
22. Solé, R.V., Ferrer-Cancho, R., Montoya, J.M., Valverde, S.: Complexity 8(1), 20–33 (2003)

# Effectiveness of Close-Loop Congestion Controls for DDoS Attacks

Takanori Komatsu and Akira Namatame

Mathematics and Computer Science, National Defence Academy of Japan
Hashirimizu 1-10-20Yokosuka-shi, Kanagawa-Pref, Japan 239-8686
{g45045,nama}@nda.ac.jp

**Summary.** High-bandwidth traffic aggregates may occur during times of flooding-based distributed denial-of-service (DDoS) attacks which are also known as flash crowds problems. Congestion control of these traffic aggregates is important to avoid congestion collapse of network services. We perform fundamental researches to minimize the effect using existing congestion controls. We simulate DDoS attacks in different Internet topologies (Tiers model, Transit-Stub model, Scale-free model). We try to improve network resistance against DDoS attacks and similar overflow problems by using open-loop and close-loop congestion controls such as Droptail, RED and CHOKe. Furthermore we propose a new congestion contorl method based on protocol type of flow and compare the performance with existing methods.

## 1 Introduction

There are various security risks in the Internet. One of these security risks, is so called DDoS atacks, which can make network congested and bring servers down with huge packets. DDoS attacks have two general forms:

1. Force the victim computer(s) to reset or consume its resources such that it can no longer provide its intended service. (For examples SYN Flood etc)
2. Obstruct the communication media between the users and the victim in such that they can no longer communicate adequately. (For examples UDP Flood etc)

There are several approaches to DDoS attacks. In this paper we researched congestion control methods based on bandwidth control against the UDP Flood problem in complex network. A UDP Flood attack is a denial-of-service (DoS) attack using User Datagram Protocol (UDP). An attack on a specific host can cause extreme network congestion in addition to hostfs performance decreasing. In the result, normal flows in the network are restricted by attack flows. Before now, there are many defenses for flooding-based DDoS attacks. While much current effort focuses on simple network topology, DDoS attack occurs in the Internet which has complex network properties. In this paper, several different Internet topologies (Tiers model [2], Transit-Stub model [9][8], Scale-free model

M. Gen et al.: Intelligent and Evolutionary Systems, SCI 187, pp. 79–90.
springerlink.com 

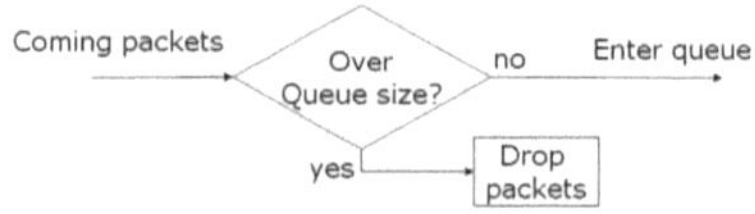


**Fig. 1.** Open-loop congestion control:Droptail

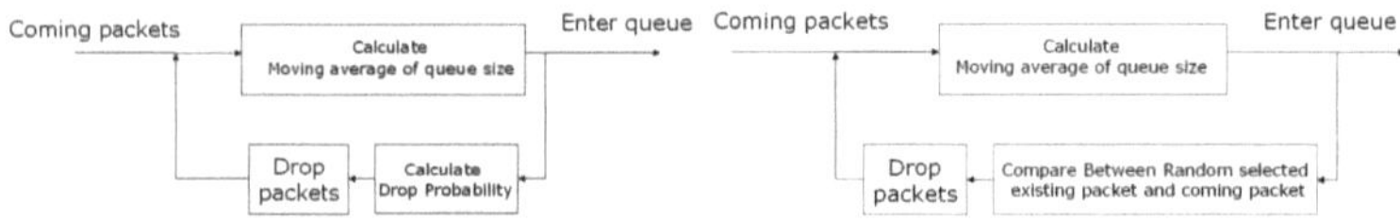


**Fig. 2.** Close-loop congestion control:RED (left) and CHOKe (right)

[1]) are uesd for simulation. These network represents each properties of real Internet Topology(Hierarchy, Domain Architectures, Scale-free etc.).

Packet filtering based on signature data base is one of major methods toward UDP flooding. The problem of the method is that it is too dificult to make complete data base. So system administrator is annoyed with false positive and false negative alarm. If false positive happens, the legitimate flows cannot pass the firewall at all. Because the bandwidth contorl based on queue management doesn't use signature data base, we can ignore the problem of false positive and negative. Therefoe we apply it to congestion problems.

We evaluate and compare three queue methods and our proposal method under congestion situation. At first, DropTail queue method is used. Droptail is classified into open loop congestion control(Fig.1). This is basic one and has first input first output queue (FIFO). Second, RED [6] and CHOKe[4] queue methods are used. RED and CHOKe queue methods are classified into close loop congestion control(Fig.2). They use feedback data about queue size information to improve their behavior against congestion problem. Finally we propose protocol based queuing methods as congestion control. In this method, packets of unselfish flow, which has contorl mechanism of sending rate like TCP, are managed by open loop congestion control and packets of selfish flow, which has no control mechanism of sending rate like UDP, are managed by closed loop congestion control.

The remainder of this paper is structured as follows. In section 2, we introduce related works. In section 3, existing congestion control methods and our proposal method are introduced. In section 4, several network topologies used in our simulation are introduced. In section 5 and 6, we explain our simulation scenario and results respectively. Finally, in section 7 we present the conclusions and future work.

## 2 Related Literature

### 2.1 CITRA

The CITRA (Cooperative Intrusion Traceback and Response Architecture) architecture [7] was designed to mitigate the effects of DoS attacks by using a rate-limiting mechanism (bandwidth control), which is quite similar to the aggregate congestion control with a pushback system presented in the next chapter.

The latest published version of CITRA has a two-level organization. At the highest level, administrative domains controlled by a component called the Discovery Coordinator (DC) are called CITRA communities. A DC is a device with human oversight that controls and monitors activity throughout a community. One community is then divided into CITRA neighborhoods. A neighborhood is a set of CITRA-enabled devices that are directly adjacent, i.e., that are not separated by any CITRA-enabled boundary controller, such as routers or firewalls. Every CITRA-enabled device collects network audit data. If one device detects an attack, it sends the attack identification data to its neighbors and requests that they check whether they are also on the attack path. Neighbors compare the attack pattern with their own audited data and determine whether they are on the attack path. If they are on the attack path, they repeat the request to their own neighbors. Thus, the attack is gradually traced back to its source or to the boundary of the CITRA system. In addition to tracing the attack, each CITRA-enabled device also performs an automated response defined according to a certain policy. Possible actions include blocking the traffic and limiting its authorized bandwidth.

The CITRA architecture is implemented and tested[7]. The tests deal only with well-identified traffic aggregates. Only the attack traffic suffered rate-limiting, while the legitimate traffic passed through the system without penalties.

However, perfect traffic aggregate identification is not currently possible. The performances of IDSs suffer from false positives. However, if perfect attack detection were possible, why would rate-limiting be used when blocking would be more effective? By filtering based on the chacharacteristics of packets (ex. source address), packets which are classfied into false positive can't pass router devices. Rate-limiting can avoid that situation. This is why we focus on rate-limiting to mitigagte DDoS attack and evaluate it.

## 3 Congestion Control Methods

Several rate-limiting congestion control methods have been proposed to mitigate internet traffic. In the present study, we used methods of the following forms.

### 3.1 Droptail

Droptail has a finite queue and implements FIFO scheduling, as shown in Fig.3. This is typical of most present-day Internet routers. Droptail is a rather simple discipline that does not rely on estimating traffic properties. If the queue is full, no incoming packets can enter the queue until the buffer space becomes available. Thus, sometimes the queue is filled by only one flow. Droptail does not have a congestion avoidance mechanism. Traffic bursts are common in packet networks, and, hence, an almost full droptail queue may cause multiple packet drops.

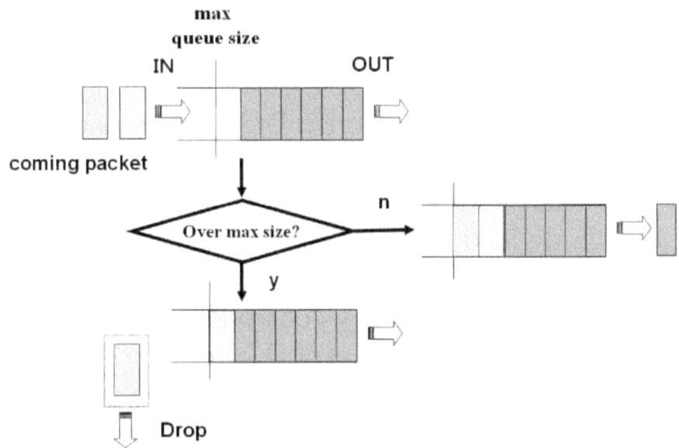


**Fig. 3.** Diagram of droptail

### 3.2 Random Early Detection: RED

RED[6] is an advanced queue method, as shown in Fig.4. RED drops packets from the queue with a certain probability, which increases with the exponential moving average queue length. Thus, the queue is not filled by only one flow (which will happen in droptail). RED does not classify traffic. Efficient packet dropping requires several configuration parameters: buffer capacity, lower threshold $min_{th}$, upper threshold $max_{th}$, and weight coefficient $w_q$. RED continuously estimates the exponential moving average queue length ($avg$) from instantaneous queue length ($q$):

$$avg_i = (1 - w_q)avg_{i-1} + w_q q \tag{1}$$

Threshold parameters $min_{th}$ and $max_{th}$ divide the buffer into three areas. The value of $avg$ controls the behavior of the RED management. No packets are discarded if avg is smaller than the ($min_{th}$) threshold. RED acts if avg is between the lower ($min_{th}$) and upper ($max_{th}$) thresholds by dropping packets with a drop probability that is linearly proportional to the exponential moving average

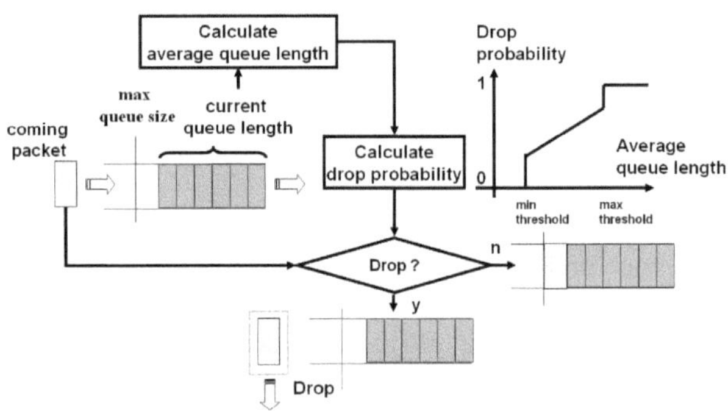


**Fig. 4.** Mechanism of RED

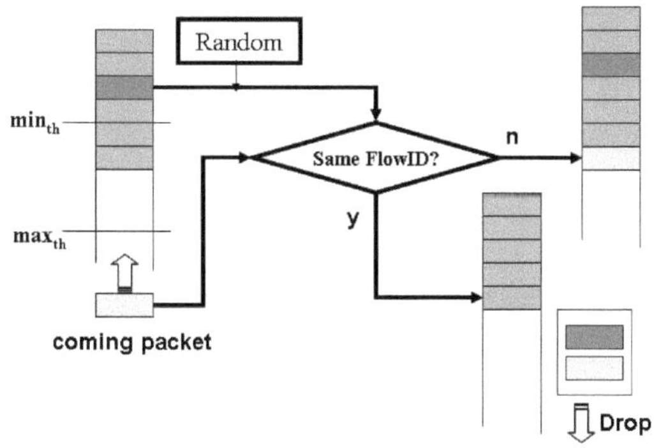


**Fig. 5.** Diagram of CHOKe

queue size. These probabilistic drops are called early drops. They serve as an indication of an imminent congestion. An optimal operation of the RED mechanism should maintain the exponential moving average queue length within the $(min_{th}, max_{th})$ area. RED functions as a droptail when the exponential moving average queue length increases beyond $max_{th}$.

### 3.3 CHOKe

CHOKe[4] has mechanism which differentially penalizes unresponsive and unfriendly flows, as shown in Fig.5.

The behavior of CHOKe is determined by two threshold values ($Min_{th}$, $Max_{th}$). If the exponential moving average queue size which is used in RED is less than $min_{th}$, each arriving packet is queued into the FIFO buffer.

If the exponential moving average queue size is larger than $min_{th}$, each arriving packet is compared with a randomly selected packet, called the drop candidate packet, from the FIFO buffer. If these packets have the same flow ID, they are both dropped (referred to herein as the preferential drop mechanism). Otherwise, the randomly chosen packet is kept in the buffer (in the same position as before), and the arriving packet is queued.

If the exponential moving average queue size is greater than $max_{th}$, each arriving packet is compared with a randomly selected packet, called the drop candidate packet, from the FIFO buffer. If these packets have the same flow ID, they are both dropped. Otherwise, the randomly chosen packet is kept in the buffer (in the same position as before) and the arriving packet is dropped. This returns the exponential moving average queue size to below $max_{th}$.

The difference between CHOKe and droptail is the use of a preferential packet drop mechanism when the exponential moving average queue size exceeds threshold and using exponential moving average queue size.

Figure 6 shows the CHOKe process in the present simulation. In this figure, "Random Packet" means that a packet which is randomly selected from the queue.

Consider two type flows (large and small) that enter the same router. If the aggregated incoming rate is smaller than the output link capacity, the queue

1. Check QueueLength.
2. if($QueueLength < Min_{th}$)
   A new packet can enter the queue.
3. if($Min_{th} \leq QueueLength \leq Max_{th}$)
   Check ($RandomPacket, ComingPacket$) Same ID?
4. if(Yes)
   $DROP(RandomPacket)$;
   $DROP(ComingPacket)$;
5. if(No)
   $Enque(ComingPacket)$;
6. $if(Max_{th} < QueueLength \leq QueueCapacity)$
   Do [Check ID Process] three times.
7. if(Not all Random Packets have the same ID as the Coming packet)
   $Enque(ComingPacket)$;
8. if($QueueLength + 1 > QueueCapacity$)
   Do [Check ID Process] three times.
   $DROP(ComingPacket)$;

**Fig. 6.** Pseudo-code of CHOKe

size does not increase to $min_{th}$. If the aggregated incoming rate is grater than the output link capacity, the queue size increases. In addition, the size of each packet depends on each flow rate. In fact, in the queue, the number of packets belonging to a large flow is larger than the number of packets belonging to a small flow. Therefore, more packets of a large flow are dropped by the process of packet comparison. This mechanism is very simple, but must be realized using a preferential drop mechanism.

### 3.4 Protocol Based Queuing Method

The precursor of the Internet, ARPANet (Advanced Research Projects Agency Network) is born in 1969[5]. And then, the Internet have grown as the system designed for the research world. Therefore the Internet has no mechanism to punish selfish (attack) flow. Basically the sending rate of each flow depends on end user behavior. This offers simple network system and help expansion of the Internet. However this permits attackers to wreak a lot of damage on the Internet performance.

The consept of protocol based queueing (PBQ) is that the network protect it's function by itself. Selfish flow should be managed by close loop congestion contorl(CLCC) and autonomous flow should be managed by open loop congestion control(OLCC). Concretely speaking, in protocol based queuing at layer 4, UDP flow is managed by RED and TCP flow is managed by droptail(Fig.7).

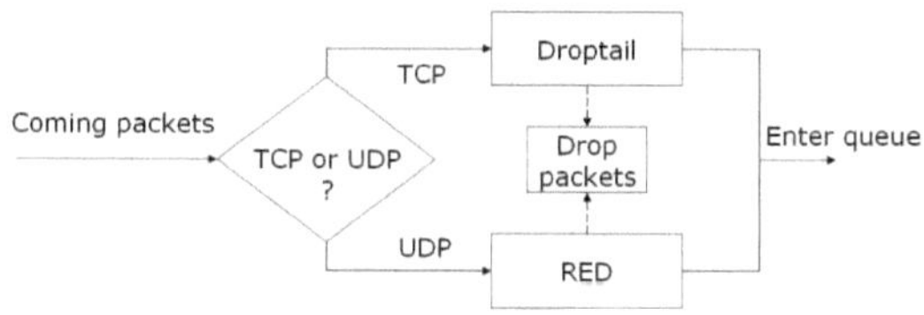


**Fig. 7.** Diagram of PBQ method

# 4 Network Topologies Used for Simulations

The real Internet is considered to consist of several topologies, depending on the point of view. We thus take into account all properties needed to simulate DDoS attacks. In this section, we discuss network topologies used to simulate DDoS attacks.

## 4.1 Tiers Model

The Internet has a hierarchical structure, as shown in Fig.8 [2]. In this model, nodes are categorized into three types: edge nodes (LAN nodes), bridge, router or switch nodes (Metropolitan Area Network - MAN nodes) and gateway (WAN) nodes. Empirically, this idea is very natural. For example, in the Science Information Network, which is the Internet Information Infrastructure for universities and research institutes in Japan, many universities connect to key university (MAN), which is connected to a backbone WAN. In addition, many university clients are connected to each other by a LAN.

## 4.2 Transit-Stub Model

At present, the Internet can be viewed as a collection of interconnected routing domains, which are groups of nodes under a common administration that share routing information. A primary characteristic of these domains is routing locality, in which the path between any two nodes in a domain remains entirely within the domain. Thus, each routing domain in the Internet can be classified as either a stub or transit domain (Fig.9).

**Fig. 8.** Tiers model

**Fig. 9.** Transit-stub model

A domain is a stub domain if the path connecting nodes u and v passes through that domain and if either u or v is located in that domain. Transit domains do not have this restriction. The purpose of transit domains is to interconnect stub domains efficiently. Without transit domains every pair of stub domains would need to be directly connected. Stub domains can be further classified as single- or multi-homed. Multi-homed stub domains have connections to more than one other domain. Single-homed stubs connect to only one transit domain. A transit domain is comprised of a set of backbone nodes, which are typically high-connected to each other.

### 4.3 Scale-Free Network (Barabasi-Albert (BA) Model)

The property of this model is that the degree distribution obeys the power law, which is observable in Internet AS level Topology. The main features with respect to how to make Barabasi-Albert (BA) model are:

1. Networks expand continuously by the addition of new nodes.
2. New nodes preferentially attach to sites that are already well connected.

## 5 Simulation Scenario

In this section, we explain how to make the simulation netowrk and traffic.

### 5.1 Network Generation

The network consists of edge (link) and node. The each edge has a buffer which store packets, which is waiting for sending to next node. And the packets in a buffer is managed by congestion control method based on queue management. The capacity of buffer is same between all edges. The number of waiting packets in a buffer keep increasing under congestion, and packets are dropped when the number exceeds the capacity of buffer. The edge has delay time. It takes a some time for packets to go through the edge. Threfore the time for packets to go through the edge is the sum of waiting time at a buffer plus delay time.

There are two type host in the network. One is host node which send, receive and route flows. Another one is router node which only route flows. In our simulation, all nodes are host nodes which consist of TCP host node and UDP host node. TCP host nodes send and receive TCP flows. And UDP host nodes send and receive UDP flows. We show parameter settings in each network in table 1.

### 5.2 Traffic Generation

There are TCP (Reno) and UDP flow in the network. Each session of flow is generated in a random manner as follows.

**Table 1.** Parameter settings in simulation network

| Network@ | Tiers | Transit-stub | BA |
|---|---|---|---|
| The number of nodes | 1000 | 1008 | 1000 |
| The number of links | 1364 | 1409 | 2994 |
| Average hop number | 16.1 | 8.7 | 3.5 |
| Diameter | 37 | 21 | 6 |
| Link bandwidth@ | 10 [Mbps] | | |
| Link delay | 10 [ms] | | |
| Queue | Droptail,RED,CHOKe | | |
| Queue size | 500 [packets] | | |

1. The number of tcp host nodes are decieded by tcp host rate $(1 - p)$ and total number of hosts $N$. Among them $N(1-p)$ tcp host nodes are deployed randomly over the network.
2. Second, each tcp host nodes select one destination host randomly. And tcp sessions are made between those host nodes. Then, there are $N(1 - p)$ tcp sessions in the network.
3. Third, same methods are done for the rest $Np$ UDP host nodes.

Next, FTP service is deployed on each TCP session. The size of data FTP service wants to transfer is infinity. The sendig rate of TCP flow is decided by TCP Reno mechanism (slow start and congestion avoidance).

And constant bit rate (CBR) service is deployed on each UDP session. That means UDP flow doesn't change it's sending rate during simulation. The sending rate of UDP flow is about twice value of average TCP throughput over 1000 TCP flows when there are only TCP flow in each network. Because UDP flow occupies twice bandwidth of TCP flow, the network becomes congestion phase as the number of UDP flow increases. We show parameter settings in each flow in table 2.

**Table 2.** Parameters in TCP and UDP flow

| @ | Network | | |
|---|---|---|---|
| TCP | Tiers | TS | Scale-free |
| The value of sendig rate | Depend on TCP Reno mechanism | | |
| UDP | Tiers | TS | Scale-free |
| The value of sendig rate [Mbps] | 0.595 | 0.730 | 4.23 |

By this means, we make TCP and UDP flows in the network. We control the amount of TCP flow and UDP flow in the network by changing variable $p \in [0, 0.9]$.

## 6 Simulation Results

In this section, we show our experiments results. TCP and UDP flows are made by changing $p$ the proportion of UDP host over 1000 nodes. Therefore the number of TCP flow is $1000(1-p)$ and the number of UDP flow is $1000p$ in each network. The throughput value in results is normalized by average TCP throughput over $1000(1-p)$ flows at $p=0$ in each network(Fig.10,11,12). We evaluate the impact of network topology and congestion contorl toward average throughput. From the results, at $p=0.1$ we can see the average throughput of UDP flow gets about two times higher than the average throughput of TCP flow by using existing congestion control methods in all network topology. Autonomous flow (TCP flow) is restricted by selfish flow (UDP flow). And average TCP throughput decreases as the proportion of UDP host increases in all network topology.

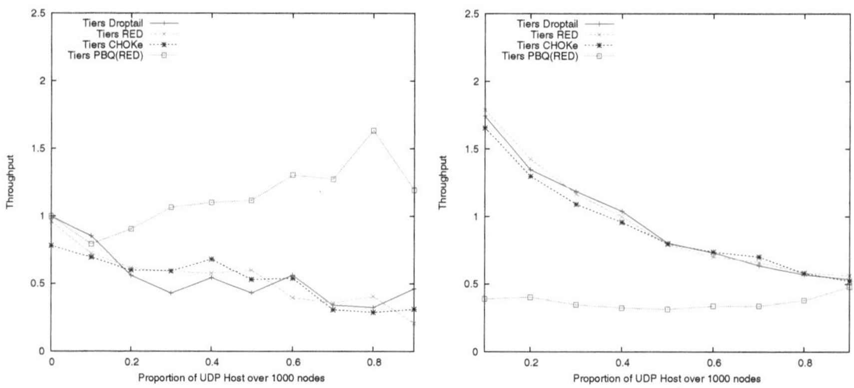


(a) TCP (b) UDP

**Fig. 10.** Average throughput per flow:Tiers model

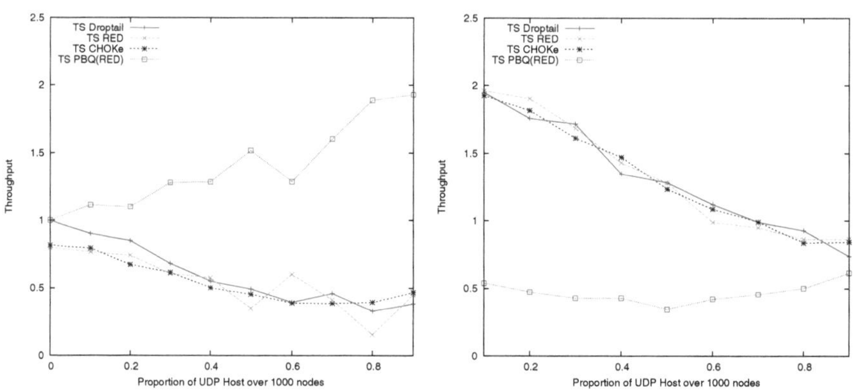


(a) TCP (b) UDP

**Fig. 11.** Average throughput per flow:Transit-stub model

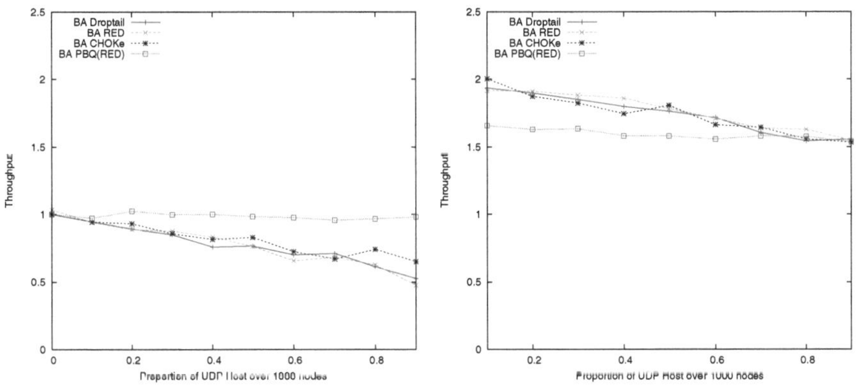


(a) TCP (b) UDP

**Fig. 12.** Average throughput per flow:Scale free network (BA model)

However PBQ moderate fluctuation of average TCP and UDP throughput well especially in BA (see Fig.12). Because the scale free BA network model has hub node. Therefore major part of flows in the nework go through a common small number of hub nodes. Congestion contorl based on queue management works effectively in that situation. These results show us that PBQ can be the way to mitigate congestion problem.

## 7 Conclusion and Future Work

Congestion control based on queue management gives simple and efficient way of managing over flow problem. Because that doesn't need signature dabase, the implementation cost becomes very low. In this paper, we propose protocol based queuing method. And we compare the effects of congestion controls in several different network topologies by numerical experiments. From the results of experiments, PBQ at layer 4 moderate average throughput fluctuation in different network topologies. That show us PBQ can be the basement measure against congestion problem.

However we classify flows by layer 4 state (TCP and UDP), there are many applications on TCP and UDP in real network. We will enhance the idea PBQ from at layer 4 state to upper layer state, and also analyze how many nodes and which nodes we have to implement PBQ method to mitigate congestion problem efficiently.

## References

1. Barabasi, A.L., Albert, R.: Emergence of scaling in random networks. Science 286, 509–512 (1999)
2. Doar, M.B.: A better model for generating test networks. In: Proceedings of global telecommunications conference, pp. 86–93 (November 1996)

3. Komatsu, T., Namatame, A.: Defending against high-bandwidth traffic aggregates. IJCSNS International Journal of Computer Science and Network Security 7(2), 243–250 (2007)
4. Pan, R., Prabhakar, B., Psounis, K.: A stateless active queue management. scheme for approximating fair bandwidth allocation. In: Proceedings of IEEE Infocom., pp. 942–951 (April 2000)
5. Remondino, F., Chen, T.-c.: Isprs and internet: History, presence and future. In: International Archives of Photogrammetry and Remote Sensing (September 2002)
6. Floyd, S., Jacobson, V.: Random early-detection gateways for congestion avoidance. IEEE/ACM Transantions on Networking 1(4), 397–413 (1993)
7. Schnackenberg, D., Holliday, H., Smith, R., Djahandari, K., Sterne, D.: Cooperative intrusion traceback and response architecture (citra). In: Proceedings of the second DARPA information survivability conference and exposition, pp. 56–58 (June 2001)
8. Zegura, E.W., Calvert, K.L., Donahoo, M.J.: A quantitative comparison of graph-based models for internet topology. IEEE/ACM Transactions on networking 5(6), 770–783 (1997)
9. Zegura, E.W., Calvert, K., Bhattacharjee, S.: How to model an internetwork. In: Proceedings of IEEE Infocom., pp. 594–602 (March 1996)

# Priority-Based Genetic Algorithm for Shortest Path Routing Problem in OSPF

Lin Lin and Mitsuo Gen

Graduate School of Information, Production and Systems, Waseda University
linlin@aoni.waseda.jp, gen@waseda.jp

**Abstract.** With the growth of the Internet, Internet service providers try to meet the increasing traffic demand with new technology and improved utilization of existing resources. Routing of data packets is the most important way to improve network utilization. Open Shortest Path First (OSPF) is the first widely deployed routing protocol that could converge a network in the low seconds, and guarantee loop-free paths. In this paper, we propose a new shortest path routing algorithm by using a priority-based Genetic Algorithm (priGA) approach in OSPF. Different with traditional Dijkstra's algorithms, GAs provide us great flexibility, robustness and adaptability to make efficient implementations for specific routing problems, such as Quality of Service (QoS) requirements, OSPF weight setting *etc*. Numerical experiments with various scales of network problems show the effectiveness and the efficiency of our approach by comparing with the recent researches.

**Keywords:** Genetic Algorithm, Open Shortest Path First, Priority-based Encoding, Shortest Path Routing.

## 1 Introduction

With the growth of the Internet, Internet Service Providers (ISPs) try to meet the increasing traffic demand with new technology and improved utilization of existing resources. Routing of data packets can affect network utilization. Packets are sent along network paths from source to destination following a protocol. Open Shortest Path First (OSPF) is the most commonly used protocol [1]. OSPF uses a Shortest Path Routing (SPR) algorithm to calculate routes in the routing table. The SPR algorithm computes the shortest (least cost) path between the router and all the networks of the internetwork. SPR routes are always calculated to be loop-free. Instead of exchanging routing table entries like Routing Information Protocol (RIP) routers, OSPF routers maintain a map of the internetwork that is updated after any change to the network topology. This map, called the link state database, is synchronized between all the OSPF routers and is used to compute the routes in the routing table. Neighboring OSPF routers form an adjacency, which is a logical relationship between routers to synchronize the link state database.

However, as the size of the link state database increases, memory requirements and route computation times increase. Furthermore, current OSPF is called "best-effort" routing protocols, which means it will try its best to forward user traffic, but can provide no guarantees regarding loss rate, bandwidth, delay, delay jitter, *etc*. For example

M. Gen et al.: Intelligent and Evolutionary Systems, SCI 187, pp. 91–103.
springerlink.com 

video-conferencing and video on-demand, which require high bandwidth, low delay, and low delay jitter. And provide the different types of network services at the same time are very difficult by using the traditional Dijkstra's algorithms.

Recently, Neural Networks (NNs), GAs, and other evolutionary algorithms have received a great deal of attention regarding their potential as optimization techniques for network design problems [2][3] and are often used to solve many real world problems: the shortest path routing (SPR) problem [4]-[6], multicasting routing problem [4], ATM bandwidth allocation problem [7], capacity and flow assignment (CFA) problem [8], and the dynamic routing problem [9]. It is noted that all these problems can be formulated as some sort of a combinatorial optimization problem.

Munemoto *et al.* propose an adaptive routing algorithm for packet-switching network such as the Internet which tries to minimize communication latency by observing delay of the routes [10]. They employ a GA based on *variable-length chromosomes* to construct a routing table that is a population of strings each of which represents a route. But the algorithm requires a relatively large population for an optimal solution due to the constraints on the crossover mechanism, and is not suitable for large networks or real-time communications.

Ahn and Ramakrishna propose a GA for solving the SPR problem. Variable-length chromosomes have been employed [5]. Their elements represent nodes included in a path between a designated pair of source and destination nodes. The crossover exchanges partial chromosomes (partial-routes) and the mutation introduces new partial chromosomes (partial-routes). Lack of positional dependency in respect of crossing sites helps maintain diversity of the population. But crossover may generate infeasible chromosomes that generating loops in the routing paths. Therefore it must be checked that none of the chromosomes is infeasible at each generation, and is not suitable for large networks or unacceptable high computational complexity for real-time communications involving rapidly changing network topologies.

Inagaki *et al.* proposed an algorithm that employs fixed (deterministic) length chromosomes [4]. The chromosomes in the algorithm are sequences of integers and each gene represents a node ID that is selected randomly from the set of nodes connected with the node corresponding to its locus number. All the chromosomes have the same (fixed) length. In the crossover phase, one of the genes (from two parent chromosomes) is selected at the locus of the starting node ID and put in the same locus of an offspring. One of the genes is then selected randomly at the locus of the previously chosen gene's number. This process is continued until the destination node is reached. The details of mutation are not explained in the algorithm. The algorithm requires a large population to attain an optimal or high quality of solution due to its inconsistent crossover mechanism. Some offspring may generate new chromosomes that resemble the initial chromosomes. Therefore we lose feasibility and heritability.

In this paper, we propose a new GA approach for solving the SPR problem in OSPF. The proposed method adopts priority-based encoding method to represent a path in the network. Numerical experiments with various scales of network problems show the effectiveness and the efficiency of our approach by comparing with the recent researches. The paper is organized as follows: In Section 2, the SPR problem is defined, and proposed GA approach with a new crossover operator, weight mapping crossover (WMX) is discussed in Section 3. Computational results including a performance comparison with the previous method are given in Section 4. Section 5 includes conclusion.

## 2 Mathematical Formulation

Let $G = (N,A)$ be a *directed network*, which consists of a finite set of nodes $N = \{1, 2, ..., n\}$ and a set of directed arcs $A=\{(i, j), (k, l), ..., (s, t)\}$ connecting $m$ pairs of nodes in $N$. Arc $(i, j)$ is said to be incident with nodes $i$ and $j$, and is directed from node $i$ to node $j$. Suppose that each arc $(i, j)$ has been assigned to a nonnegative value $c_{ij}$, the cost of $(i, j)$. The SRP can be defined by the following assumptions:

A1. The network is directed. We can fulfil this assumption by transforming any undirected network into a directed one.
A2. All transmission delay and all arc costs are nonnegative.
A3. The network does not contain parallel arcs (i.e., two or more arcs with the same tail and head nodes). This assumption is essentially for notational convenience.

**Indices**

$i, j, k$: index of node $(1, 2, ..., n)$

**Parameters**

$n$: number of nodes

$c_{ij}$: transmission cost of arc $(i, j)$

**Decision variables**

$x_{ij}$: the link on an arc $(i, j)$

The SPP is to find the minimum cost z from a specified source node 1 to another specified sink node $n$, which can be formulated as follows in the form of integer programming:

$$\min \quad z = \sum_{i=1}^{n} \sum_{j=1}^{n} c_{ij} x_{ij} \tag{1}$$

$$\text{s. t.} \quad \sum_{j=1}^{n} x_{ij} - \sum_{k=1}^{n} x_{ki} = \begin{cases} 1 & (i = 1) \\ 0 & (i = 2, 3, \cdots, n-1) \\ -1 & (i = n) \end{cases} \tag{2}$$

$$x_{ij} = \{0, 1\}, \qquad \forall (i, j) \tag{3}$$

## 3 Genetic Approach for SPR Problem

### 3.1 Genetic Representation

How to encode a solution of the problem into a chromosome is a key issue for GAs. For any application case, it is necessary to perform analysis carefully to ensure an appropriate representation of solutions together with meaningful and problem-specific genetic operators [3]. One of the basic features of GAs is that they work on coding space and solution space alternatively: genetic operations work on coding space (chromosomes), while evaluation and selection work on solution space. For the nonstring coding approach, three critical issues emerged concerning with the encoding and decoding between chromosomes and solutions (or the mapping between phenotype and genotype): (1) The feasibility of a chromosome; (2) The legality of a chromosome; (3) The uniqueness of mapping.

Feasibility refers to the phenomenon of whether a solution decoded from a chromosome lies in the feasible region of a given problem. Legality refers to the phenomenon of whether a chromosome represents a solution to a given problem. The illegality of chromosomes originates from the nature of encoding techniques. For many combinatorial optimization problems, problem-specific encodings are used and such encodings usually yield to illegal offspring by a simple one-cut-point crossover operation. Because an illegal chromosome cannot be decoded to a solution, it means that such chromosomes cannot be evaluated. Repairing techniques are usually adopted to convert an illegal chromosome to a legal one. The mapping from chromosomes to solutions (decoding) may belong to one of the following three cases: 1-to-1 mapping, $n$-to-1 mapping and 1-to-$n$ mapping. The 1-to-1 mapping is the best one among three cases and 1-to-$n$ mapping is the most undesired one. We need to consider these problems carefully when designing a new non-binary-string coding so as to build an effective GA.

Gen *et al.* proposed priority-based encoding firstly for the solving shortest path problem [6]. In this paper, we extend the priority-based encoding method. As is known, a gene in a chromosome is characterized by two factors: locus, *i.e.*, the position of the gene located within the structure of chromosome, and allele, *i.e.*, the value the gene takes. In this encoding method, the position of a gene is used to represent node ID and its value is used to represent the priority of the node for constructing a path among candidates. A path can be uniquely determined from this encoding.

Illustration of priority-based chromosome and its decoded path are shown in Fig. 2, in terms of a undirected network in Fig. 1. At the beginning, we try to find a node for the position next to source node 1. Nodes 2, 3 and 4 are eligible for the position, which can be easily fixed according to adjacent relation among nodes. The priorities of them are 1, 6 and 4, respectively. Node 3 has the highest priority and is put into the path. The possible nodes next to node 3 are nodes 4 and 6. Because node 4 has the

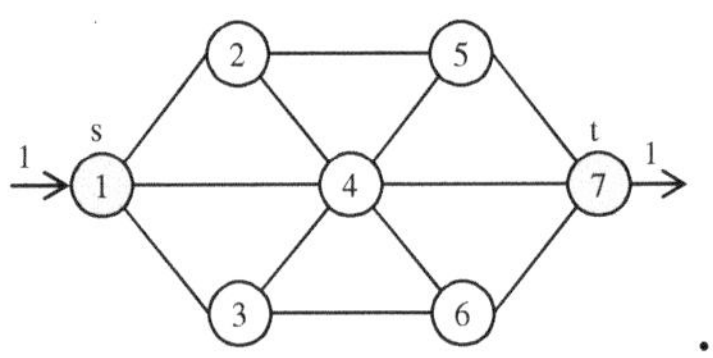


**Fig. 1.** A simple undirected graph with 7 nodes and 12 edges

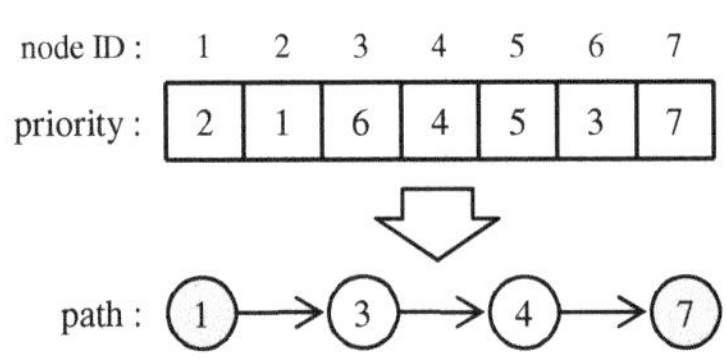


**Fig. 2.** An example of priority-based chromosome and its decoded path

largest priority value, it is put into the path. Then we form the set of nodes available for next position and select node 7 with the highest priority among them. Repeat these steps until we obtain a complete path, (1-3-4-7). The encoding process and decoding process of the priority-based GA (priGA) are shown in Figures 3 and 4, respectively.

```
procedure 1: priority - based encoding method
input : number of nodes n
output : kth initial chromosome v_k
begin
    for i = 1 to n
        v_k[i] ← i ;
    for i = 1 to ⌈n / 2⌉
        repeat
            j ← random[1,n];
            l ← random[1,n];
        until(j ≠ l);
        swap(v_k[j], v_k[l]);
    output v_k;
end
```

**Fig. 3.** Pseudocode of priority-based encoding method

```
procedure 2: path growth
input: number of nodes n, chromosome v_k, the set of nodes S_i
        with all nodes adjacent to node i
output: path P_k
begin
    initialize i ← 1, l ← 1, P_k[l] ← i;
                                // i : source node, l : length of path P_k.
    while S_1 ≠ φ do
        j' ← argmax{v_k[j], j ∈ S_i};
                    // j': the node with highest priority among S_i.
        if v_k[j'] ≠ 0 then
            P_k[l] ← j';   // chosen node j' to construct path P_k.
            l ← l + 1;
            v_k[j'] ← 0;
            i ← j';
        else
            S_i ← S_i \ j';   // delete the node l adjacent to node i.
            v_k[i] ← 0;
            l ← l − 1;
            if l ≤ 1 then l ← 1, break;
            i ← P_k[l];
    output path P_k
end
```

**Fig. 4.** Pseudocode of priority-based decoding method

The advantages of the priority-based encoding method are: (1) any permutation of the encoding corresponds to a path (feasibility); (2) most existing genetic operators can be easily applied to the encoding; (3) any path has a corresponding encoding (legality); (4) any point in solution space is accessible for genetic search. However, there is a disadvantage as that $n$-to-1 mapping (uniqueness) may occur for the encoding at some case. For example, we can obtain the same path, (1-3-4-7) by different chromosomes, ($v_1 = [2, \underline{1}, 6, 4, \underline{5}, 3, 7]$ and $v_2 = [2, \underline{5}, 6, 4, \underline{1}, 3, 7]$).

### 3.2 Fitness Function

The fitness function interprets the chromosome in terms of physical representation and evaluates its fitness based on traits of being desired in the solution. The fitness function in the SPR problem is obvious because the shortest path computation amounts to finding the minimal cost path. Therefore, the fitness function that involves computational efficiency and accuracy (of the fitness measurement) is defined as follows:

$$eval_k = \frac{1}{\sum_{l_i \in P_k} c_{l_{i-1}l_i} x_{l_{i-1}l_i}} \tag{4}$$

where $eval_k$ represents the fitness value of the $k$th chromosome, link $(l_{i-1}, l_i)$ is included in the routing path $P_k$.

### 3.3 Genetic Operators

Genetic operators mimic the process of heredity of genes to create new offspring at each generation. Using the different genetic operators has very large influence on GA performance. Therefore it is important to examined different genetic operators.

#### 3.3.1 Crossover

For priority-based representation as a permutation representation, several crossover operators have been proposed, such as partial-mapped crossover (PMX), order crossover (OX), cycle crossover (CX), position-based crossover (PX), heuristic crossover, *etc* [3].

In all the above crossover operators, the mechanism of the crossover is not the same as that of the conventional one-cut point crossover. Some offspring may be generated that did not succeed on the character of the parents, thereby the crossover retard the process of evolution. In this paper, we propose a weight mapping crossover (WMX); it can be viewed as an extension of one-cut point crossover for permutation representation. At one-cut point crossover, two chromosomes (parents) would choose a random-cut point and generate the offspring by using a segment of its own parent to the left of the cut point, then remap the right segment based on the weight of other parent of right segment. Fig. 5 shows the crossover process of WMX, and an example of the WMX is given in Fig. 6.

```
procedure 3: weight mapping crossover (WMX)
input: two parents v1, v2, the length of chromosome n
output: offspring v1', v2'
begin
  p ← random[1, n];  // p: a random cut point
  l ← n − p;  // l: the length of right segments of chromosomes
  v1' ← v1[1 : p] // v2[p + 1 : n];
  v2' ← v2[1 : p] // v1[p + 1 : n];
              // exchange substrings between parents
  s1[·] ← sorting(v1[p + 1 : n]);
  s2[·] ← sorting(v2[p + 1 : n]);
              // sorting the weight of the right segments
  for i = 1 to l
    for j = 1 to l
      if v1'[p + i] = s2[j] then v1'[p + i] ← s1[j];
    for j = 1 to l
      if v2'[p + i] = s1[j] then v2'[p + i] ← s2[j];
  output offspring v1', v2';
end
```

**Fig. 5.** Pseudocode of weight mapping crossover

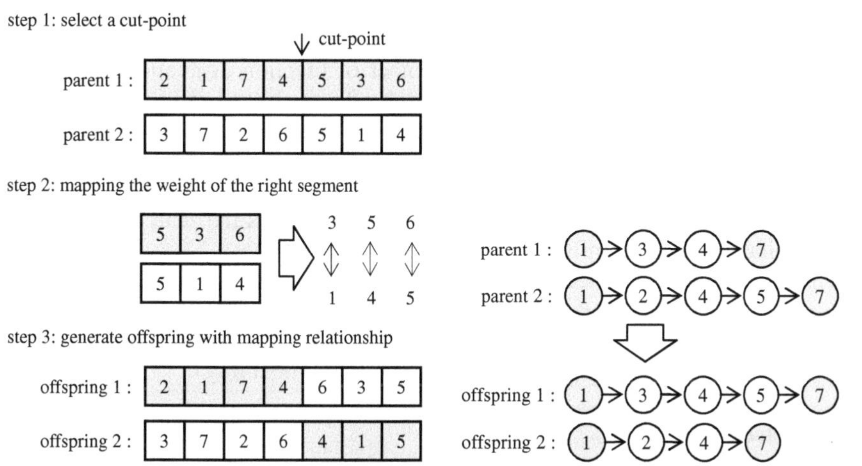


(a) Illustration of WMX

(b) Illustration of the generated paths for the undirected graph shown in Fig. 1

**Fig. 6.** An example of WMX procedure

As showed in Fig. 6, first we choose a random cut point $p$; and calculate $l$ that is the length of right segments of chromosomes, where $n$ is the number of nodes in the network. Then we get a mapping relationship by sorting the weight of the right segments $s_1[.]$ and $s_2[.]$. As a one-cut point crossover, it generates the offspring $v_1[.]$, $v_2[.]$ by exchange substrings between parents $v_1[.]$, $v_2[.]$; legalized offspring with mapping relationship and then two new chromosomes are produced eventually. For solving the SPR problem by priGA, WMX is similar with the conventional one-cut point crossover, and it can generate the offspring that exchanged sub-route from two parents.

### 3.3.2 Mutation

For permutation representation, it is relatively easy to produce some mutation operators. Several mutation operators have been proposed for permutation representation, such as swap mutation, inversion mutation, insertion mutation, *etc.*

As it is described above, $n$-to-1 mapping may occur for the priority-based encoding, in above example of the reciprocal exchange mutation procedure, the generated offspring is different as parent, but the decoded path is same. Thus it is important to improve effectiveness of mutation. In this paper, we examines several kinds of mutation operators, effectiveness of the insertion mutation is the best mutation for priority-based representation. Insertion mutation selects an element at random and inserts it in a random position.

### 3.3.3 Immigration

The trade-off between exploration and exploitation in serial GAs for function optimization is a fundamental issue. If a GA is biased towards exploitation, highly fit members are repeatedly selected for recombination. Although this quickly promotes better members, the population can prematurely converge to a local optimum of the function. On the other hand, if a GA is biased towards exploration, large numbers of schemata are sampled which tends to inhibit premature convergence. Unfortunately, excessive exploration results in a large number of function evaluations, and defaults to random search in the worst case. To search effectively and efficiently, a GA must maintain a balance between these two opposing forces. Moed *et al.* proposed an immigration operator which, for certain types of functions, allows increased exploration while maintaining nearly the same level of exploitation for the given population size [11]. It is an example of a random strategy which explores the search space ignoring the exploitation of the promising regions of the search space.

The algorithm is modified to (1) include immigration routine, in each generation, (2) generate and (3) evaluate *popSize* $p_I$ random members, and (4) replace the *popSize* $p_I$ worst members of the population with the *popSize* $p_I$ random members (*popsize* called the population size, $p_I$, called the immigration probability).

### 3.3.4 Selection

We adopt the roulette wheel selection (RWS). It is to determine selection probability or survival probability for each chromosome proportional to the fitness value. A model of the roulette wheel can be made displaying these probabilities.

## 4 Experiments and Discussion

Usually during the GA design phase, we are only concerned with the design of genetic representations, neglecting the design of more effective genetic operators with depend on the characteristics of the genetic representations. In the experiment, the effectiveness of different genetic operators will be demonstrated. Then to validate the effectiveness of different genetic representations, priority-based GA with Ahn and Ramakrishna's algorithm [5] are compared. For each algorithm, 20 runs with Java are performed on a Pentium 4 processor (3.40-GHz clock), 3.00GA RAM.

### 4.1 Test Problems

For examining the effect of different encoding methods, we applied Ahn and Ramakrishna's algorithm (Ahn's Alg.) and priority-based encoding method on 6 test problems [5][12]. Dijskstra's algorithm has been used to obtain optimal solutions for the problems and the solution qualities of the proposed priGA and Ahn's Alg. are investigated using optimal solution. Each algorithm was run 20 times using different initial seeds for each test problems. Two different stopping criteria are used. One of them is number of maximum generations. But, if the algorithm didn't improve the best solution in successive 100 generations, it is stopped to save computation time.

### 4.2 Performance Comparisons with Different Genetic Operators

In the first experiment, the different genetic operators for priority-based genetic representation are combined; there are partial-mapped crossover (PMX), order crossover (OX), position-based crossover (PX), swap mutation, insertion mutation and immigration operator. There are seven kinds of combinations of genetic operators: PMX+Swap (Alg.1), OX+Swap (Alg.2), PX+Swap (Alg.3), WMX+Swap (Alg.4), WMX+Swap+ Immigration (Alg.5) and WMX+Insertion+Immigration (Alg.6). The GA parameter settings are shown as follows:

Population size: *popSize* =20;
Crossover probability: $p_C$ =0.70;
Mutation probability: $p_M$ =0.50;
Immigration rate: $p_I$ = 0.15;
Maximum generation: *maxGen* =1000;
Terminating condition: 100 generations with same fitness.

Table 1 gives average cost of 20 run for each combination of crossover and mutation operator. The path optimality is defined in all test problems, by Alg.6

**Table 1.** Performance comparisons with different genetic operators

| Test Problems (# of nodes/ # of arcs) | Optimal Solutions | Best Solutions | | | | | |
|---|---|---|---|---|---|---|---|
| | | Alg. 1 | Alg. 2 | Alg. 3 | Alg. 4 | Alg. 5 | Alg. 6 |
| 20/49 | 142.00 | 148.35 | 148.53 | 147.70 | 143.93 | 142.00 | 142.00 |
| 80/120 | 389.00 | 423.53 | 425.33 | 418.82 | 396.52 | 389.00 | 389.00 |
| 80/632 | 291.00 | 320.06 | 311.04 | 320.15 | 297.21 | 291.62 | 291.00 |
| 160/2544 | 284.00 | 429.55 | 454.98 | 480.19 | 382.48 | 284.69 | 284.00 |
| 320/1845 | 394.00 | 754.94 | 786.08 | 906.18 | 629.81 | 395.01 | 394.00 |
| 320/10208 | 288.00 | 794.26 | 732.72 | 819.85 | 552.71 | 331.09 | 288.00 |

Alg. 1: PMX+Swap; Alg. 2: OX+Swap; Alg. 3: PX+Swap; Alg. 4: WMX+Swap;
Alg. 5: WMX+Swap+Immigration(3); Alg. 6: WMX+Insertion+Immigration(3).

(WMX+Insertion+Immigration) that the GA finds the global optimum (*i.e.*, the shortest path). The path optimality is defined in 1st, 2nd test problems, by Alg.5 (WMX+Swap+Immigration), The near optimal result is defined in other test problems. By Alg.1 ~ Alg.4, the path optimality is not defined. Since the number of possible alternatives become to very large in test problems, the population be prematurely converged to a local optimum of the function.

### 4.3 Comparisons with Different Encoding Methods

How to encode a solution of the problem into a chromosome is a key issue in GAs. The different chromosome representations will have a very big impact on GA designs. In the second experiment, the performance comparisons between priority-based GA (priGA) and Ahn and Ramakrishna's algorithm (Ahn's Alg.) are showed. In priGA, WMX crossover, insertion mutation and immigration (Alg.6) are used as genetic operators.

Table 2 gives computational results for two different encoding methods on six test problems. When we compare columns of the best cost of three encoding methods, it is possible to see that priGA developed gives better performance than Ahn's Alg.. For conceding the terminating condition (100 generations with same fitness), Ahn's Alg. cannot improve its best result in successive 100 runs for the large scale network problems, it is stopped with short computation time. Sometimes, Ahn's Alg. is faster than proposed algorithm; however its result is worse than proposed algorithm.

**Table 2.** Performance comparisons with Ahn's Alg. and proposed priGA

| Test Problems (# of nodes/ # of arcs) | Optimal Solutions | Best Solutions | | CPU Times (*ms*) | | Generation Num. of Obtained best result | |
|---|---|---|---|---|---|---|---|
| | | Alg. 6 | Ahn's Alg. | Alg. 6 | Ahn's Alg. | Alg. 6 | Ahn's Alg. |
| 20/49 | 142.00 | 142.00 | 142.00 | 23.37 | 40.60 | 9 | 2 |
| 80/120 | 389.00 | 389.00 | 389.00 | 96.80 | 118.50 | 4 | 4 |
| 80/632 | 291.00 | 291.00 | 291.00 | 118.50 | 109.50 | 10 | 19 |
| 160/2544 | 284.00 | 284.00 | 286.20 | 490.50 | 336.20 | 26 | 31 |
| 320/1845 | 394.00 | 394.00 | 403.40 | 1062.50 | 779.80 | 11 | 44 |
| 320/10208 | 288.00 | 288.00 | 288.90 | 1498.50 | 1028.30 | 26 | 38 |

### 4.4 Comparisons with Different GA Parameter Settings

In addition, as we all know, in general the result of GA decrease with increasing the problem size means that we must increase the GA parameter settings. Therefore if the parameter setting of a GA approach does not increase the problem size, then we can say this GA approach has a very good search capability of obtaining optimal results. The effectiveness comparing with six kinds of different GA parameter settings are shown as follows:

Population size: *popSize* =10, 20, 30, 50 or 100;
Crossover probability: $p_C$ =0.30 or 0.70 ;
Mutation probability: $p_M$ =0.10 or 0.50;

The quality of solution for each GA is investigated in Table 3. We can see that many factors such as population size, crossover possible, and mutation possible *etc.* can have a significant impact and their interrelationships should help in identifying the important factors and their ideal combinations for effective performance in different settings. As depicted in Table 3, Ahn's Alg. can solve the first four test problems successfully, but for solving last two test problems, GA parameter setting affected the efficiency of ahnGA.

To see clearly the difference between proposed priGA and Ahn's Alg. with the different GA parameter settings, Fig. 7 show the percent deviation from optimal solution of the 6th test problem. The values of percent deviation from optimal solution are showed in Table 4. As depicted in Fig. 7 and Table 4, if the GA parameter settings: *popSize*=20; $p_C = 0.70$ and $p_M = 0.50$ are combined, proposed priGA with all the test problems successfully solved.

**Table 3.** Performance comparisons with different parameter settings

| Parameter Settings ( $pop\_size$ / $p_C$ : $p_M$ ) | Test Problems (# of nodes/ # of arcs) | Optimal Solutions | Best Solutions | | CPU Times ( *ms* ) | | Generation Num. of Obtained best result | |
|---|---|---|---|---|---|---|---|---|
| | | | Alg. 6 | Ahn's Alg. | Alg. 6 | Ahn's Alg. | Alg. 6 | Ahn's Alg. |
| 10 / 0.3 : 0.1 | 20/49 | 142.00 | 142.00 | 156.20 | 8.37 | 10.42 | 27 | 38 |
| | 80/120 | 389.00 | 389.00 | 389.00 | 31.10 | 32.80 | 1 | 5 |
| | 80/632 | 291.00 | 291.00 | 313.20 | 34.40 | 29.40 | 16 | 43 |
| | 160/2544 | 284.00 | 284.20 | 320.90 | 106.30 | 67.10 | 37 | 48 |
| | 320/1845 | 394.00 | 394.00 | 478.70 | 250.20 | 120.30 | 18 | 68 |
| | 320/10208 | 288.00 | 288.30 | 444.00 | 400.20 | 126.40 | 59 | 25 |
| 20 / 0.3 : 0.1 | 20/49 | 142.00 | 142.00 | 145.23 | 13.34 | 22.36 | 24 | 27 |
| | 80/120 | 389.00 | 389.00 | 389.00 | 51.50 | 56.30 | 1 | 4 |
| | 80/632 | 291.00 | 291.00 | 303.10 | 56.30 | 50.10 | 10 | 18 |
| | 160/2544 | 284.00 | 284.20 | 298.70 | 181.20 | 122.10 | 35 | 44 |
| | 320/1845 | 394.00 | 394.00 | 465.70 | 496.70 | 213.90 | 17 | 32 |
| | 320/10208 | 288.00 | 288.60 | 373.10 | 631.10 | 311.00 | 35 | 61 |
| 20 / 0.7 : 0.5 | 20/49 | 142.00 | 142.00 | 142.00 | 23.37 | 40.60 | 9 | 6 |
| | 80/120 | 389.00 | 389.00 | 389.00 | 96.80 | 118.50 | 1 | 1 |
| | 80/632 | 291.00 | 291.00 | 291.00 | 118.50 | 109.50 | 10 | 19 |
| | 160/2544 | 284.00 | 284.00 | 286.20 | 490.50 | 336.20 | 26 | 31 |
| | 320/1845 | 394.00 | 394.00 | 403.40 | 1062.50 | 779.80 | 11 | 44 |
| | 320/10208 | 288.00 | 288.00 | 288.90 | 1498.50 | 1028.30 | 26 | 38 |

**Table 4.** Percent deviation from optimal solution for the 6th problem

| Parameter Settings ( $pop\_size$ / $p_C$ : $p_M$ ) | Probability of obtaining the optimal solutions | |
|---|---|---|
| | Prop. Alg. | Ahn's Alg. |
| 10 / 0.3 : 0.1 | 66.67% | 16.67% |
| 20 / 0.3 : 0.1 | 66.67% | 16.67% |
| 30 / 0.3 : 0.1 | 83.33% | 33.33% |
| 50 / 0.3 : 0.1 | 100.00% | 50.00% |
| 100 / 0.3 : 0.1 | 100.00% | 33.33% |
| 10 / 0.7 : 0.5 | 83.33% | 33.33% |
| 20 / 0.7 : 0.5 | 100.00% | 50.00% |
| 30 / 0.7 : 0.5 | 100.00% | 50.00% |
| 50 / 0.7 : 0.5 | 100.00% | 83.33% |
| 100 / 0.7 : 0.5 | 100.00% | 83.33% |

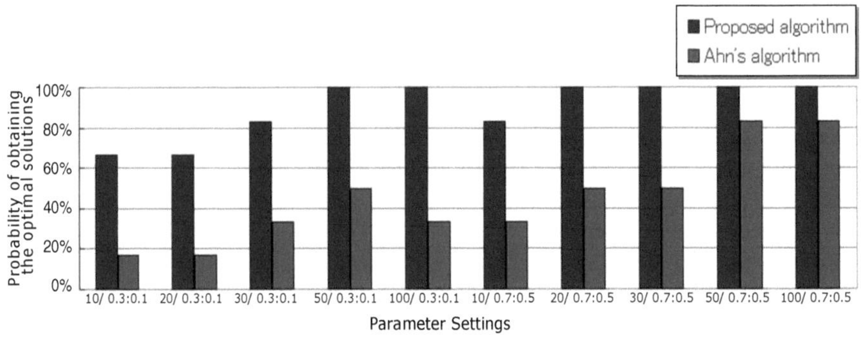


**Fig. 7.** Performance comparisons by percent deviation from optimal solution for the 6th problem

# 5 Conclusion

In this paper, we proposed a GA approach to the Shortest Path Routing (SPR) problem in OSPF. The proposed method adopted a priority-based encoding method to represent a path in the network. We also proposed a new crossover operator, *weight mapping crossover* and adopted *insertion mutation* and *immigration operator* in this paper. Numerical experiment showed the efficiency and effectiveness of the GA approach on the SPR problem. In the experimental study, a preliminary study was realized to investigate the effects of different combinations of crossover and mutation operators on the performance of the priority-based encoding method, and showed computational results for two different encoding methods on six test problems. Lastly we showed that population size, crossover probability, mutation probability *etc.* can have a significant impact for effective performance in different settings. Numerical experiment demonstrated that the proposed GA approach in this paper gave better performance than the recent research on the SPR problem.

**Acknowledgments.** This work is partly supported by the Ministry of Education, Science and Culture, the Japanese Government: Grant-in-Aid for Scientific Research (No.19700071).

# References

1. Ericsson, M., Resende, M.G.C., Pardalos, P.M.: A Genetic Algorithm for the Weight Setting Problem in OSPF Routing. Journal of Combinatorial Optimization 6(3), 299–333 (2002)
2. Gen, M., Cheng, R.: Evolutionary Network Design: Hybrid Genetic Algorithms Approach. International Journal of Computational Intelligence and Applications 3(4), 357–380 (2008)
3. Gen, M., Cheng, R., Lin, L.: Network Models and Optimization: Multiobjective Genetic Algorithm Approach. Springer, Heidelberg (2008)
4. Inagaki, J., Haseyama, M., Kitajim, H.: A genetic algorithm for determining multiple routes and its applications. In: Proceeding of IEEE International Symposium. Circuits and Systems, pp. 137–140 (1999)

5. Ahn, C.W., Ramakrishna, R.: A genetic algorithm for shortest path routing problem and the sizing of populations. IEEE Transactions on Evolutionary Computation 6(6), 566–579 (2002)
6. Gen, M., Cheng, R., Wang, D.: Genetic algorithms for solving shortest path problems. In: Proceeding of IEEE International Conference on Evolutionary Computer, pp. 401–406 (1999)
7. Bazaraa, M., Jarvis, J., Sherali, H.: Linear Programming and Network Flows, 2nd edn. John Wiley & Sons, New York (1990)
8. Mostafa, M.E., Eid, S.M.A.: A genetic algorithm for joint optimization of capacity and flow assignment in Packet Switched Networks. In: Proceeding of 17th National Radio Science Conference, pp. C5-1–C5-6 (2000)
9. Shimamoto, N., Hiramatsu, A., Yamasaki, K.: A Dynamic Routing Control Based On a Genetic Algorithm. In: Proceeding of IEEE International Conference Neural Networks, pp. 1123–1128 (1993)
10. Munetomo, M., Takai, Y., Sato, Y.: An Adaptive Network Routing Algorithm Employing Path Genetic Operators. In: Proceeding of the Seventh Inter. Conference on Genetic Algorithms, pp. 643–649 (1997)
11. Michael, C.M., Stewart, C.V., Kelly, R.B.: Reducing the Search Time of A Steady State Genetic Algorithm using the Immigration Operator. In: Proceeding of IEEE International Conference on Tools for AI San Jose, CA, pp. 500–501 (1991)
12. OR-Notes (accessed), `http://people.brunel.ac.uk/~mastjjb/jeb/info.html`

# Evolutionary Network Design by Multiobjective Hybrid Genetic Algorithm

Mitsuo Gen[1], Lin Lin[1], and Jung-Bok Jo[2]

[1] Graduate School of Information, Production and Systems, Waseda University
gen@waseda.jp, linlin@aoni.waseda.jp
[2] Division of Computer and Information Engineering, Dongseo University
Phone: +81-90-9565-2964
jobok@gdsu.dongseo.ac.kr

**Abstract.** Network design is one of the most important and most frequently encountered classes of optimization problems. It is a combinatory field in combinatorial optimization and graph theory. When considering a bicriteria network design (bND) problem with the two conflicting objectives of minimizing cost and maximizing flow, network design problems where even one flow measure be maximized, are often NP-hard problems. But, in real-life applications, it is often the case that the network to be built is required to optimize multi-criteria simultaneously. Thus the calculation of the multi-criteria network design problems is a difficult task. In this paper, we propose a new multiobjective hybrid genetic algorithm (mo-hGA) hybridized with Fuzzy Logic Control (FLC) and Local Search (LS). Numerical experiments show the effectiveness and the efficiency of our approach by comparing with the recent researches.

**Keywords:** Genetic Algorithm, Priority-based Encoding, Fuzzy Logic Control, Local Search, Bicriteria Network Design.

## 1 Introduction

Network design is one of the most important and most frequently encountered classes of optimization problems. It is a combinatory field in combinatorial optimization and graph theory. A lot of optimization problems in network design arose directly from everyday practice in engineering and management: determining shortest or most reliable paths in traffic or communication networks, maximal or compatible flows, or shortest tours; planning connections in traffic networks; coordinating projects; and solving supply and demand problems. Furthermore, network design is also important for complexity theory, an area in the common intersection of mathematics and theoretical computer science which deals with the analysis of algorithms [1]. However, there is a large class of network optimization problems for which no reasonable fast algorithms have been developed. And many of these network optimization problems arise frequently in applications. Given such a hard network optimization problem, it is often possible to find an efficient algorithm whose solution is approximately optimal. Among such techniques, genetic algorithm (GA) is one of the most powerful and broadly applicable stochastic search and optimization techniques based on principles from evolution theory.

Network design problem couples deep intellectual content with a remarkable range of applicability, covering literally thousands of applications in such wide-ranging

M. Gen et al.: Intelligent and Evolutionary Systems, SCI 187, pp. 105–121.
springerlink.com 

fields as chemistry and physics, computer networking, most branches of engineering, manufacturing, public policy and social systems, scheduling and routing, telecommunications, and transportation.

Shortest path problem (SPP), maximum flow problem (MXF) and minimum cost flow problem (MCF) *etc.* are also well known basic network design problems. While in SPP, a path is determined between two specified nodes of a network that has minimum length, or the maximum reliability or takes least time to traverse, MXF finds a solution that sends the maximum amount of flow from a source node to a sink node. MCF is the most fundamental of all network design problems. In this problem, the purpose is to determine a least cost shipment of a commodity through a network in order to satisfy demands at certain nodes from available supplies at other nodes (Ahuja, 1993). These problems have been well studied and many efficient polynomial-time algorithms have been developed by Dijsktra (1959), Dantzig (1960), Ford and Fulkerson (1956), Elias *et. al.* (1956), Ford and Fulkerson (1962) and Zadeh (1973) [2].

In many applications, network design is often the case that the network to be built is required to optimize multi-criteria simultaneously. The problems may arise when designing communication networks, manufacturing systems, and logistic systems. For example, in a communication network, find a set of links which consider the low cost (or delay) and the high throughput (or reliability) for increasing the network performance (e.g., [3, 4]); in a manufacturing system, the two criteria under consideration are minimizing cost and maximizing manufacturing [5]; or in a logistic system, the main drive to improve logistics productivity is the enhancement of customer services and asset utilization through a significant reduction in order cycle time (lead time) and logistics costs [6].

The Bicriteria Network Design (bND) problem is known as NP-hard [7], it is not simply an extension from single objective to two objectives. In generally, we cannot get the optimal solution of the problem because these objectives usually conflict with each other in practice. The real solutions to the problem are a set of Pareto optimal solutions [8]. For solving the bND problem, the set of efficient paths may be very large and possibly exponential in size. Thus the computational effort required to solve it can increase exponentially with the problem size in the worst case. While the tractability of the problem is of importance when solving large scale problems, the issue concerning with the size of the efficient set is important to a decision maker. Having to evaluate a large efficient set in order to select the best one poses a considerable cognitive burden on decision makers. Therefore, in such cases, obtaining the entire Pareto optimal set is of little interest to decision makers.

The bicriteria shortest path problem is one of bND problems, which of finding a diameter constrained shortest path from a specified source node *s* to another specified sink node *t*. This problem, termed the multi-objective shortest path problem (moSPP) in the literature, is NP-hard and Warburton (1987) presented the first fully polynomial approximation scheme (FPAS) for it [9]. Hassin (1992) provided a strongly polynomial FPAS for the problem which improved the running time of Warburton [10].

The general classes of bND problems with minimum two objectives (under different cost functions) are defined and extended to the more multi-criteria network design problems. Ravi *et al.* (1994) presented an approximation algorithm for finding good broadcast networks [11]. Ganley *et al.* (1995) consider a more general problem with more than two objective functions [12]. Marathe *et al.* (1998) consider three different criteria of network and presented the first polynomial-time approximation algorithms for a large class of bND problem [2].

In this research, we considered bND problem with more complexity case as two criteria functions that maximize total flow and minimize total cost considered. Where even one flow measure is maximized, are often NP-hard [13], because of its several unique characteristics. For example, a flow at each edge can be anywhere between zero and its flow capacity, *i.e.,* it has more "freedom" to choose. In many other problems, selecting an edge may mean to simply add fixed distance. It has been well studied using a variety of methods by parallel algorithm with a worst case time of $O(n^2 \log n)$ (Shiloach and Vishkin, 1982), distributed algorithms with a worst case time of $O(n^2 \log n)$ to $O(n^3)$ (Yeh and Munakata, 1986), and recent sequential algorithms etc., with $n$ nodes. But the computational effort required to solve it can increase with the problem size.

GA has received considerable attention regarding their potential as a novel optimization technique. There are three major advantages when applying GA to optimization problems:

1. *Adaptability:* GA does not have much mathematical requirements about the optimization problems. Due to the evolutionary nature, GA will search for solutions without regard to the specific inner workings of the problem. GA can handle any kind of objective functions and any kind of constraints, i.e., linear or nonlinear, defined on discrete, continuous or mixed search spaces.
2. *Robustness:* The use of evolution operators makes GA very effective in performing global search (in probability), while most of conventional heuristics usually perform local search. It has been proved by many studies that GA is more efficient and more robust in locating optimal solution and reducing computational effort than other conventional heuristics.
3. *Flexibility:* GA provides us a great flexibility to hybridize with domain-dependent heuristics to make an efficient implementation for a specific problem.

Multiple objective problems arise in the design, modeling, and planning of many complex real systems in the areas of industrial production, urban transportation, capital budgeting, forest management, reservoir management, layout and landscaping of new cities, energy distribution, etc. It is easy to find that almost every important real world decision problem involves multiple and conflicting objectives which need to be tackled while respecting various constraints, leading to overwhelming problem complexity. The multiple objective optimization problems have been receiving growing interest from researchers with various background since early 1960 [14].

The inherent characteristics of the GA demonstrate why genetic search is possibly well suited to multiple objective optimization problems. The basic feature of the GA is the multiple directional and global searches by maintaining a population of potential solutions from generation to generation. The population-to-population approach is useful to explore all Pareto solutions. The GA does not have much mathematical requirement regarding the problems and can handle any kind of objective functions and constraints. Due to their evolutionary nature, the GA can search for solutions without regard to the specific inner workings of the problem. Therefore, it is applicable to solving much more complex problems beyond the scope of conventional methods' interesting by using the GA. Because the GA, as a kind of meta-heuristics, provides us a great flexibility to hybridize with conventional methods into their main framework, we can take with advantage of both the GA and the conventional methods to make much more efficient implementations for the problems. The ongoing research on applying the GA to the multiple objective optimization problems presents a formidable theoretical and practical challenge to the mathematical community [15].

For applying a GA approach for this complexity case of bND problem, *priority-based encoding method* has been improved. For maximizing total flow, different form general genetic representation methods, such as path oriented encoding method, priority-based encoding method can represent various efficient paths by each chromosome. Considering the characteristic of priority-based encoding method, we propose a new crossover operator called as *Weight Mapping Crossover* (WMX), *Insertion mutation operator* is adopted. These methods provide a search capability that results in improved quality of solution and enhanced rate of convergence. For ensure the population diversity in multiobjective GA, *interactive Adaptive-weight Genetic Algorithm* (i-awGA) which is one of weighted-sum fitness assignment approach, is improved. Their elements represent that assign weights to each objective and combines the weighted objectives into a single objective function. Weights are adjusted adaptively based on the current generation to obtain search pressure toward the positive ideal point. The rest of the paper is organized as follows: The rest of the paper is organized as follows: In Sect. 2, we formulate the mathematic model of bND problem. In Sect. 3, we propose a hybrid genetic algorithm (hGA) with combining fuzzy logic control (FLC) and local search (LC). This hGA obtain an effective implementation of GAs to network models and real applications. We propose a priority-based encoding method, a weight mapping crossover (WMX) and an immigration operator for the network problems. We also propose an interactive adaptive-weight fitness approach for multi-criteria network problems. In Sect. 4, we demonstrate effectiveness comparing with different combination of genetic operators and also demonstrate effectiveness comparing with different fitness assignment approaches by moGAs. Finally, we give the conclusion follows in Sect. 5.

## 2 Mathematical Formulation

Consider a directed network $G = (N, A)$, consisting of a finite set of nodes $N = \{1, 2, \ldots, n\}$ and a set of directed arcs $A = \{(i, j), (k, l), \ldots, (s, t)\}$ joining pairs of nodes in $N$. Arc $(i, j)$ is said to be incident with nodes $i$ and $j$, and is directed from node $i$ to node $j$. We shall assume that the network has $n$ nodes and $m$ arcs. Fig. 1 presents a simple network with 11 nodes and 22 arcs.

We associate a total flow $f$ that is the available supply of an item $f$ or the required demand for the item $-f$. Node with $f$ is called source node and node s (or 1) with $-f$ is called sink node t (or $n$). If $f$=0, node $i$ is called an intermediate (or transshipment) node. Associated with each arc $(i, j)$ has a *capacity* $u_{ij}$ that denotes the maximum amount of flow on the arc $(i, j)$ and a *lower bound* 0 that denotes the minimum amount of flow, we associate $x_{ij}$ be the amount of flow $(0 \le x_{ij} \le u_{ij})$ and $c_{ij}$ be the unit shipping cost along the arc $(i, j)$. We consider the Bicriteria Network Design (bND) subject to the following assumptions.

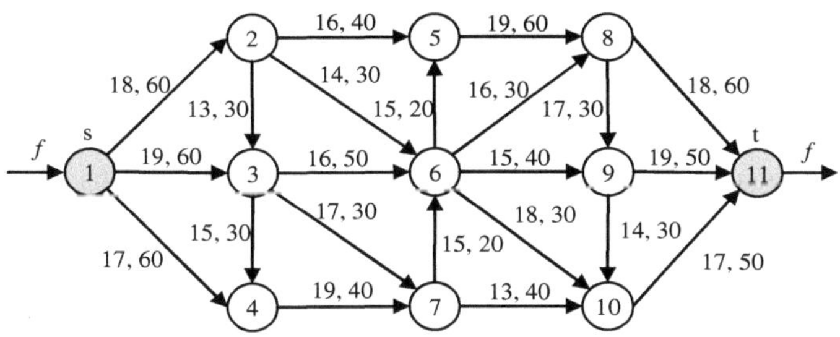


**Fig. 1.** A simple network with 11 nodes and 22 edges

**A1.** The network is directed. We can fulfill this assumption by transforming any undirected network into a directed network.

**A2.** All capacities are nonnegative integers. Although it is possible to relax the integrality assumption on arc capacities for some algorithms, this assumption is necessary for others. Algorithms whose complexity bounds involve $U$ assume integrality of the data. In reality, the integrality assumption is not a restrictive assumption because all modern computers store capacities as rational numbers and we can always transform rational numbers to integer numbers by multiplying them by a suitably large number.

**A3.** The network does not contain a directed path from node s to node t composed only of infinite capacity arcs. Whenever every arc on a directed path $P$ from note s to note t has infinite capacity, we can send an infinite amount of flow along this path, and therefore the maximum flow value is unbounded.

**A4.** The network does not contain parallel arcs (i.e., two or more arcs with the same tail and head nodes). This assumption is essentially a notational convenience.

The decision variables in the bND problem are the maximum possible flow $z_1$ with minimum cost $z_2$ from source node 1 to sink node $n$. Mathematically, this problem formulated as follows (where summations are taken over existing arcs):

$$\max \quad z_1 = f \tag{1}$$

$$\min \quad z_2 = \sum_{i=1}^{n}\sum_{j=1}^{n} c_{ij} x_{ij} \tag{2}$$

$$\text{s. t.} \quad \sum_{k=1}^{n} x_{jk} - \sum_{i=1}^{n} x_{ij} = \begin{cases} f & (j=1) \\ 0 & (j=2,3,\cdots,n-1) \\ -f & (j=n) \end{cases} \tag{3}$$

$$0 \le x_{ij} \le u_{ij}, \qquad \forall (i,j) \tag{4}$$

$$f \ge 0 \tag{5}$$

Constraints (3) are called the flow conservation or Kirchhoff equations and indicate that the flow may be neither created nor destroyed in the network. In the conservation equations, sum of $x_{jk}$ represents the total flow out of node $j$ while sum of $x_{ij}$ indicates the total flow into node $j$. These equations require that the net flow out of node $j$, should equal $f$. If $f<0$, then there should be more flow into $j$ than out of $j$.

# 3 Multiobjective Hybrid Genetic Algorithm

## 3.1 Genetic Representation

For many real-world applications, it is nearly impossible to represent their solutions with the binary encoding. Various encoding methods have been created for particular problems in order to have an effective implementation of the GA. In this paper, we consider a priority-based encoding method with special decoding for various network design problems.

Given a new encoding method, it is usually necessary to examine whether we can build an effective genetic search with the encoding. Several principles have been proposed to evaluate an encoding [15]:

Property 1 (*Space*): Chromosomes should not require extravagant amounts of memory.
Property 2 (*Time*): The time complexity of executing evaluation, recombination and mutation on chromosomes should not be a higher order.
Property 3 (*Feasibility*): A chromosome corresponds to a feasible solution.
Property 4 (*Legality*): Any permutation of a chromosome corresponds to a solution.
Property 5 (*Completeness*): Any solution has a corresponding chromosome.
Property 6 (*Uniqueness*): The mapping from chromosomes to solutions (decoding) may belong to one of the following three cases: 1-to-1 *mapping*, *n*-to-1 *mapping* and 1-to-*n mapping*. The 1-to-1 mapping is the best among three cases and 1-to-*n* mapping is the most undesir one.
Property 7 (*Heritability*): Offspring of simple crossover (*i.e.*, one-cut point crossover) should correspond to solutions which combine the basic feature of their parents.
Property 8 (*Locality*): A small change in chromosome should imply a small change in its corresponding solution.

Cheng and Gen proposed priority-based Genetic Algorithm (priGA) firstly for solving Resource-constrained Project Scheduling Problem (rcPSP) [16]. Gen *et al.* [17] and Lin *et al.* [18] also adopted this method for solving SPP problem. The priority-based encoding method is an indirect approach. As it is known, a gene in a chromosome is characterized by two factors: locus, *i.e.*, the position of gene located within the structure of chromosome, and allele, *i.e.*, the value the gene takes. An example of priority-based encoding is shown in Fig 2. In this encoding method, the position of a gene is used to represent task ID and its value is used to represent the priority of the task for constructing assignment sequence among candidates. In this paper, we developed priGA with different decoding procedure to considering the characteristic of these network design problems.

| | | | | | | | | | | | |
|---|---|---|---|---|---|---|---|---|---|---|---|
| locus : | 1 | 2 | 3 | 4 | 5 | 6 | 7 | 8 | 9 | 10 | 11 |
| node ID : | 11 | 1 | 10 | 3 | 8 | 9 | 5 | 7 | 4 | 2 | 6 |

**Fig. 2.** An example of priority-based chromosome

To describe this decoding method, we first present a *one-path growth procedure* that obtains a path base on the generated chromosome with given network. We then present an *overall-path growth procedure* that remove the used flow from each arc and delete the arcs which its capacity is 0. Based on updated network, we obtain a new path with one-path growth procedure, and repeat these steps until obtain overall possible path.

### 3.1.1 One-Path Growth Procedure

The path is generated by one-path growth procedure that is given in procedure 2; with beginning from the specified node 1 and terminating at the specified node $n$. At each step, there are usually several nodes available for consideration. We add the one with the highest priority into path. As shown in Fig. 3, a path is decoded by the generated chromosome with the given network.

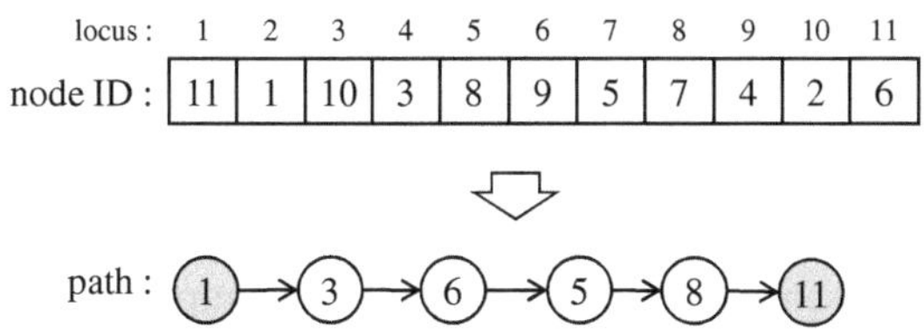


**Fig. 3.** Example of generated chromosome and its decoded path

Illustration of priority-based chromosome and its decoded path are shown in Fig. 3. At the beginning, we try to find a node for the position next to source node 1. Nodes 2, 3 and 4 are eligible for the position, which can be easily fixed according to adjacent relation among nodes. The priorities of them are 1, 10 and 3, respectively. Node 3 has the highest priority and is put into the path. The possible nodes next to node 3 are nodes 4, 6 and 7. Because node 6 has the largest priority value, it is put into the path. Then we form the set of nodes available for next position and select the one with the highest priority among them. Repeat these steps until we obtain a complete path, (1-3-6-5-8-11).

The advantages of the priority-based encoding method are: (1) the length of the code equal the number of nodes, smaller than any other encoding method in the network models (space); (2) any permutation of the encoding corresponds to a solution (feasibility); (3) any solution of network models has a corresponding the code by a priority-based encoding (legality, completeness); (4) most existing genetic operators can be easily applied to the encoding (heritability, locality).

### 3.1.2 Overall-Path Growth Procedure

For a given path, we can calculate its flow $f_k$ and the cost $c_k$. By subtracting the used capacity from $u_{ij}$ of each arc, we have a new network with the new flow capacity $\tilde{u}_{ij}$.

With the one-path growth procedure, we can obtain next path. By repeating this procedure we can obtain the maximum flow for the given chromosome until no new network can be defined in this way.

## 3.2 Genetic Operators

Genetic operators mimic the process of heredity of genes to create new offspring at each generation. Using the different genetic operators has very large influence on GA performance. Therefore it is important to examined different genetic operators.

### 3.2.1 Crossover

For priority-based representation as a permutation representation, several crossover operators have been proposed, such as partial-mapped crossover (PMX), order crossover (OX), cycle crossover (CX), position-based crossover (PX), heuristic crossover, *etc* [15]. In all the above crossover operators, the mechanism of the crossover is not the same as that of the conventional one-cut point crossover. Some offspring may be generated that did not succeed on the character of the parents, thereby the crossover retard the process of evolution. In this paper, we propose a weight mapping crossover (WMX); it can be viewed as an extension of one-cut point crossover for permutation representation. At one-cut point crossover, two chromosomes (parents) would choose a random-cut point and generate the offspring by using a segment of its own parent to the left of the cut point, then remap the right segment based on the weight of other parent of right segment. An example of the WMX is given in Fig. 4.

### 3.2.2 Mutation

For permutation representation, it is relatively easy to produce some mutation operators. Several mutation operators have been proposed for permutation representation, such as swap mutation, inversion mutation, insertion mutation, *etc.*

In this paper, we examines several kinds of mutation operators, effectiveness of the insertion mutation is the best mutation for priority-based representation. Insertion mutation selects an element at random and inserts it in a random position.

### 3.2.3 Selection

In this research, for solving bND problems by the weighted-sum approach, the selection operator is mainly used to adjust genetic search in favor of a wide exploration of the search space. We adopted roulette wheel selection that is the best selection type as supplementary to the weighted-sum approach.

## 3.3 Hybridization

GA has proved to be a versatile and effective approach for solving optimization problems. Nevertheless, there are many situations in which the simple GA does not perform particularly well, and various methods of *hybridization* have been proposed. One of most common forms of *hybrid genetic algorithm* (hGA) is to incorporate local optimization as an add-on extra to the canonical GA loop of recombination and selection. With the hybrid approach, *local optimization* is applied to each newly generated

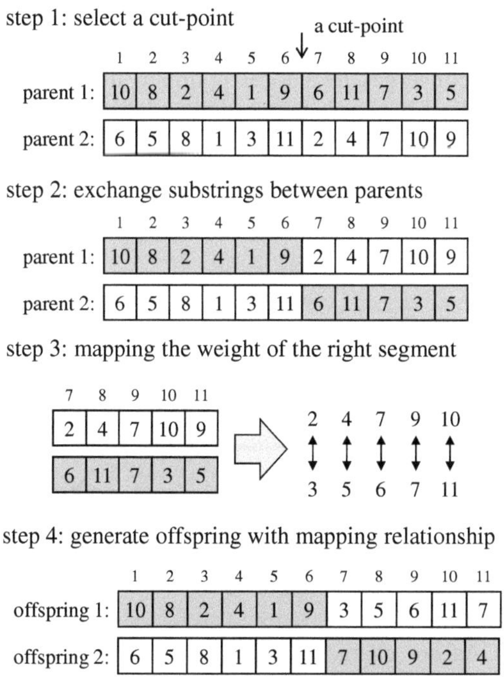


**Fig. 4.** An example of WMX

offspring to move it to a local optimum before injecting it into the population. GA is used to perform *global exploration* among a population while heuristic methods are used to perform *local exploitation* around chromosomes. Because of the complementary properties of GA and conventional heuristics, the hybrid approach often outperforms either method operating alone. Another common form is to incorporate GA parameters adaptation. The behaviors of GA are characterized by the balance between exploitation and exploration in the search space. The balance is strongly affected by the strategy parameters such as *population size, maximum generation, crossover probability*, and *mutation probability*. How to choose a value to each of the parameters and how to find the values efficiently are very important and promising areas of research on the GA.

### 3.3.1 Hybridization with Local Search

The idea of combining GA and local search heuristics for solving optimization problems has been extensively investigated and various methods of hybridization have been proposed. Applying a local search technique to GA loop is shown in Fig. 5. There are two common form of genetic local search. One featurs *Lamarckian evolution* and the other featurs the *Baldwin effect* [19]. Both approaches use the metaphor that an individual learns (hillclimbs) during its lifetime (generation). In the Lamarckian case, the resulting individual (after hillclimbing) is put back into the population. In the Baldwinian case, only the fitness is changed and the genotype remains unchanged.

According to Whitley, Gordon and Mathias' experiences on some test problems, the Baldwinian search strategy can sometimes converge to a global optimum when the Lamarckian strategy converges to a local optimum using the same form of local search. However, in all of the cases they examined, the Baldwinian strategy is much slower than the Lamarckian strategy.

In this research, we combine a Lamarckian case. The LS is applied to each newly generation, select the best individual and adopt Insertion Mutation until the best offspring is generated, and injecting it into the population.

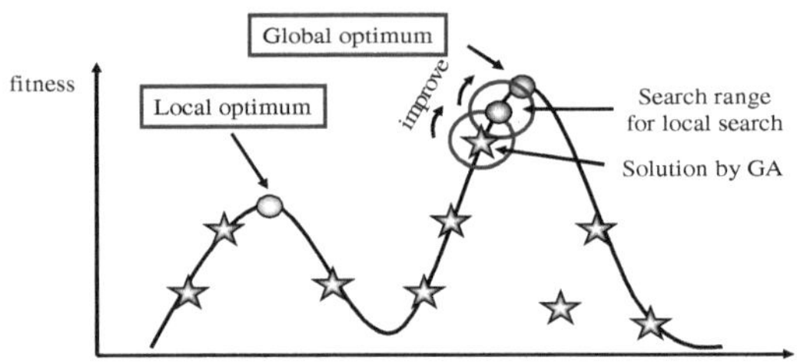


**Fig. 5.** Applying a local search technique to GA loop

### 3.3.2 Parameter Adaptation

The behaviors of GAs are characterized by the balance between exploitation and exploration in the search space. The balance is strongly affected by the strategy parameters such as *population size*, *maximum generation*, *crossover probability*, and *mutation probability*. How to choose a value to each of the parameters and how to find the values efficiently are very important and promising areas of research of the GAs.

Usually, fixed parameters are used in most applications of the GAs. The values for the parameters are determined with a set-and-test approach. Since an GA is an intrinsically dynamic and adaptive process, the use of constant parameters is thus in contrast to the general evolutionary spirit. Therefore, it is a natural idea to try to modify the values of strategy parameters during the run of the algorithm. It is possible to do this in various ways:

1. by using some rule;
2. by taking feedback information from the current state of search;
3. by employing some self-adaptive mechanism.

In our implementation of fuzzy logic control (FLC), we modify the Wang, *et al.'s,* 1997 concept to automatically regulate the probability of crossover and mutation during the evolutionary process. Let $\Delta f(t)$ is the different of average fitness function between the $t$th and ($t$-1)th generation, In this approach, the inputs of FLC are $\Delta f(t)$ and $\Delta f(t+1)$. The outputs of the FLC are $\Delta c(t)$ and$\Delta m(t)$ (the change of crossover probability and mutation probability respectively). The membership functions of fuzzy all input and output linguistic variables are illustrated in Fig. 6 [20].

Based on a number of experimental data and domain expert opinion, the input values are respectively normalized into integer values in the range [-4.0, 4.0] according to their corresponding maximum/minimum values. The control action value for the

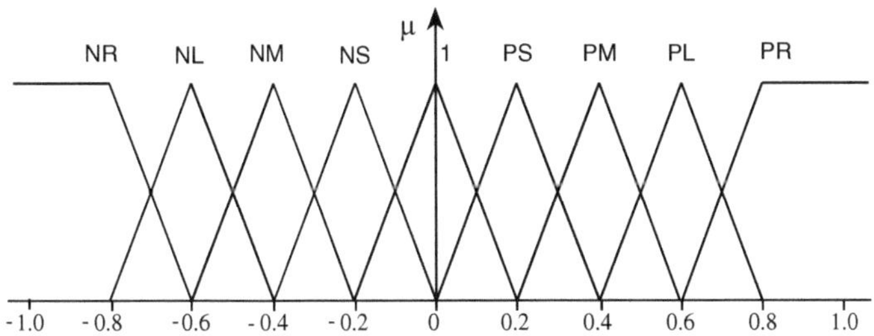


where NR – Negative larger, NL – Negative large, NM – Negative medium
NS – Negative small, ZE – Zero, PS – Positive small
PM – Positive medium, PL – Positive large, PR – Positive larger

**Fig. 6.** The membership function for input-output FLC variables

crossover and mutation operations is determined by using a look-up table as given in Table 1. Then, the changes on crossover and mutation ratios are determined as follows:

$$\Delta c(t)=0.02z(i, j),\ \Delta m(t)=0.002z(i, j),$$
$$\text{where } i,j \in \{-4, -3, -2, -1, 0, 1, 2, 3, 4\}.$$

The value of crossover ratio and mutation ratio for the next generation is calculated as follows:

$$p_C(t)= p_C(t\text{-}1)+\Delta c(t), \qquad p_M(t)= p_M(t\text{-}1)+ \Delta m(t)$$

**Table 1.** Control Action for Crossover and Mutation Ratios

| z(i, j) | | i | | | | | | | | |
|---|---|---|---|---|---|---|---|---|---|---|
| | | -4 | -3 | -2 | -1 | 0 | 1 | 2 | 3 | 4 |
| j | -4 | -4 | -3 | -3 | -2 | -2 | -1 | -1 | 0 | 0 |
| | -3 | -3 | -3 | -2 | -2 | -1 | -1 | 0 | 0 | 1 |
| | -2 | -3 | -2 | -2 | -1 | -1 | 0 | 0 | 1 | 1 |
| | -1 | -2 | -2 | -1 | -1 | 0 | 0 | 1 | 1 | 2 |
| | 0 | -2 | -1 | -1 | 0 | 2 | 1 | 1 | 2 | 2 |
| | 1 | -1 | -1 | 0 | 0 | 1 | 1 | 2 | 2 | 3 |
| | 2 | -1 | 0 | 0 | 1 | 1 | 2 | 2 | 3 | 3 |
| | 3 | 0 | 0 | 1 | 1 | 2 | 2 | 3 | 3 | 4 |
| | 4 | 0 | 1 | 1 | 2 | 2 | 3 | 3 | 4 | 4 |

### 3.3.3 Interactive Adaptive-Weight Approach

GAs are essentially a kind of meta-strategy methods. When applying the GAs to solve a given problem, it is necessary to refine upon each of the major components of GAs, such as *encoding methods*, *recombination operators*, *fitness assignment*, *selection operators*, *constraints handling*, and so on, in order to obtain a best solution to the given problem. Because the multiobjective optimization problems are the natural extensions of constrained and combinatorial optimization problems, so many useful methods based on GAs developed during the past two decades. One of special issues in the multiobjective optimization problems is fitness assignment mechanism. Since the

1980s, several fitness assignment mechanisms have been proposed and applied in multiobjective optimization problems. There are four famous fitness assignment mechanisms: random-weight Genetic Algorithm (rwGA) [21], adaptive-weight Genetic Algorithm (awGA) [20], strength Pareto Evolutionary Algorithm (spEA) [22], non-dominated sorting Genetic Algorithm II (nsGA II) [23].

Most fitness assignment mechanisms can be classified into Pareto ranking-based fitness assignment and weighted sum-based assignment. Generally, the main idea of Pareto ranking-based approach is a clear classification between nondominated solution and dominated solution for each chromosome. However, it is difficult to clear the difference among non-dominated solutions (or dominated solutions). Shown in Fig. 7(a), although there are distinct differences between the dominated solutions (2, 2) and (8, 8), but there are not distinct differences between their fitness values 13/5 and 11/5 by spEA.

Different from Pareto ranking-based fitness assignment, weighted-sum based fitness assignment is to assign weights to each objective function and combines the weighted objectives into a single objective function. It is easier to calculate the weight-sum based fitness and the sorting process becomes unnecessary, thus it is effective for considering the computation time to solve the problems. In addition, another characteristic of weighted-sum approach is use to adjust genetic search toward the Pareto frontier. For combining weighted objectives into a single objective function, the good fitness values are assigned which solutions near from the Pareto frontier. However, shown in Fig. 7(c), the fitness values (12/11, 14/11, 13/11) of most dominated solutions ((4, 10), (8, 8), (9, 6)) are greater than the fitness values (11/11, 11/11) of some non-dominated solutions ((12, 1), (1, 12)) by using awGA.

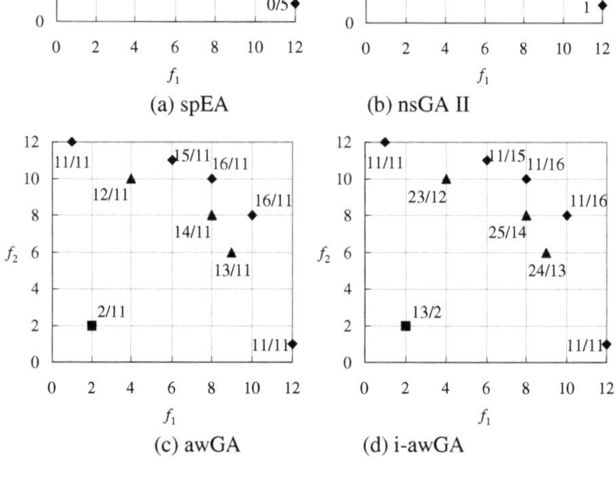


**Fig. 7.** Illustration of fitness values by different fitness assignment mechanisms

In this paper, we propose an interactive adaptive-weight fitness assignment approach (i-awGA), which is an improved adaptive-weight fitness assignment approach with the consideration of the disadvantages of weighted-sum approach and Pareto ranking-based approach. We combine a penalty term to the fitness value for all of dominated solutions. First, we define the fitness values $f_1, f_2$ as the objectives $z_1, z_2$ for two objective maximum problem Then we define two extreme points: the maximum extreme point $f^+ \leftarrow \{f_1^{max}, f_2^{max}\}$ and the minimum extreme point $f^- \leftarrow \{f_1^{min}, f_2^{min}\}$ in the current population, where $f_1^{max}$ and $f_2^{max}$ are maximum values of objective 1 and 2, respectively, $f_1^{min}$ and $f_2^{min}$ are minimum values of objective 1 and 2, respectively. Calculate the adaptive weight $w_1 = 1/(f_1^{max} - f_1^{min})$ for objective 1 and the adaptive weight $w_2 = 1/(f_2^{max} - f_2^{min})$ for objective 2. Afterwards, calculate the penalty term $p(\boldsymbol{v}_k)$=0, if $\boldsymbol{v}_k$ is nondominated solution in the nondominated set $P$. Otherwise $p(\boldsymbol{v}_{k'})$=1 for dominated solution $\boldsymbol{v}_{k'}$. Last, calculate the fitness value of each chromosome by combining the method as follows and we adopted roulette wheel selection as supplementary to the i-awGA.

$$eval(v_k) = w_1\left(f_1^k - f_1^{min}\right) + w_2\left(f_2^k - f_2^{min}\right) + p\left(v_k\right), \quad \forall k \in popSize$$

### 3.3.4 Overall Procedure of Hybrid Genetic Algorithm

Let $P(t)$ and $C(t)$ be parents and offspring in current generation $t$. The general structure of *hybrid genetic algorithm* (hGA) is described as follows:

**procedure**: mo-hGA
**input**: problem data, GA parameters
**output**: Pareto optimal solutions $E$
**begin**
  $t \leftarrow 0$;
  initialize $P(t)$ by priority-based encoding routine;
  calculate objectives $f_i(P)$, $i = 1, ..., q$ by priority-based decoding routine;
  create Pareto $E(P)$ by nondominated routine;
  evaluate $eval(P)$ by i-awGA;
  **while** (**not** terminating condition) **do**
    create $C(t)$ from $P(t)$ by crossover routine;
    create $C(t)$ from $P(t)$ by mutation routine;
    calculate objectives $f_i(C)$, $i = 1, ... ,q$ by priority-based decoding routine;
    update Pareto $E(P,C)$ by nondominated routine;
    climb $C(t)$ by local search routine;
    evaluate $eval(P,C)$ by i-awGA;
    parameter-tuning $p_M$, $p_C$ by FLC;
    select $P(t+1)$ from $P(t)$ and $C(t)$ by selection routine;
    $t \leftarrow t+1$;
  **end**
  **output** Pareto optimal solutions $E(P,C)$
**end**

## 4 Experiments and Discussion

In this section, we show performance comparisons of multiobjective GAs for solving bND problems by different fitness assignment approaches, there are spEA, nsGA II, rwGA and i-awGA. Two maximum flow test problems are presented by Munakata and Hashier (1993), and randomly assigned unit shipping costs along each arc. In each GA approach, priority-based encoding was used, and WMX and Insertion were used as genetic operators. Each simulation was run 10 times. GA parameter settings were taken as follows: population size, *popSize* =20; crossover probability, $p_C$=0.70; mutation probability, $p_M$ =0.50.

***Performance Measures:*** Reference solution set S* of each test problem was found using all algorithms which be used in computational experiments. Each algorithm was applied to each test problem with much longer computation time and larger memory storage than the other computational experiments. Generally, we used the very large parameter specifications in all algorithms for finding the reference solution set of each test problem. We chose only nondominated solutions as reference solutions by 10 runs of the two algorithms for each test problem.

a. The number of obtained solutions $|S_j|$.

b. The ratio of nondominated solutions $R_{NDS}(S_j)$: A straightforward performance measure of the solution set $S_j$ with respect to the $J$ solution sets is the ratio of solutions in $S_j$ that are not dominated by any other solutions in $S$. The $R_{NDS}(S_j)$ measure can be written as follows:

$$R_{NDS}(S_j) = \frac{\left|S_j - \{x \in S_j \,|\, \exists r \in S^*: r \prec x\}\right|}{\left|S_j\right|}$$

c. The distance $D1_R$ measure can be written as follows:

$$D1_R = \frac{1}{|S^*|} \sum_{r \in S^*} \min\{d_{rx} \,|\, x \in S_j\}$$

where $S^*$ is a reference solution set for evaluation the solution set $S_j$; $d_{xr}$ is the distance between a current solution $x$ and a reference solution $r$.

$$d_{rx} = \sqrt{(f_1(r) - f_1(x))^2 + (f_2(r) - f_2(x))^2}$$

***Discussion of the Results:*** We compare i-awGA with spEA, nsGA II and rwGA trough computational experiments under the same stopping condition (*i.e.*, evaluation of 5000 solutions). Each algorithm was applied to each test problem 10 times and gives the average results of the 3 performance measures. In Tables 2 and 3, better results of $|S_j|$ and $D1_R$ were obtained from the i-awGA than other fitness assignment approach. The results of $R_{NDS}(S_j)$ are no large differences among the 4 fitness assignment approaches.

**Table 2.** Performance Evaluation of Fitness Assignment Approaches for the 25-node/49-arc Test Problem

| # of | $\|S_j\|$ | | | | $R_{NDS}(S_j)$ | | | | $D1_R(S_j)$ | | | |
|---|---|---|---|---|---|---|---|---|---|---|---|---|
| eval. solut. | spEA | nsGAII | rwGA | i-awGA | spEA | nsGAII | rwGA | i-awGA | spEA | nsGAII | rwGA | i-awGA |
| 50 | 41.60 | 40.60 | 40.30 | 42.40 | 0.44 | 0.42 | 0.45 | 0.49 | 201.25 | 210.63 | 205.03 | 184.12 |
| 500 | 51.40 | 56.30 | 49.40 | 54.60 | 0.54 | 0.60 | 0.53 | 0.64 | 151.82 | 124.81 | 149.44 | 132.93 |
| 2000 | 58.20 | 60.60 | 54.30 | 59.20 | 0.62 | 0.71 | 0.62 | 0.75 | 108.49 | 101.45 | 127.39 | 88.99 |
| 5000 | 60.70 | 61.60 | 58.30 | 61.40 | 0.72 | 0.80 | 0.67 | 0.82 | 79.91 | 80.70 | 103.70 | 67.14 |

**Table 3.** Performance Evaluation of Fitness Assignment Approaches for the 25-node/56-arc Test Problem

| # of | $\|S_j\|$ | | | | $R_{NDS}(S_j)$ | | | | $D1_R(S_j)$ | | | |
|---|---|---|---|---|---|---|---|---|---|---|---|---|
| eval. solut. | spEA | nsGAII | rwGA | i-awGA | spEA | nsGAII | rwGA | i-awGA | spEA | nsGAII | rwGA | i-awGA |
| 50 | 41.20 | 43.60 | 42.60 | 44.00 | 0.35 | 0.33 | 0.34 | 0.33 | 181.69 | 180.64 | 168.73 | 168.96 |
| 500 | 49.80 | 56.60 | 51.60 | 57.50 | 0.47 | 0.50 | 0.42 | 0.46 | 104.77 | 114.62 | 119.53 | 103.13 |
| 2000 | 62.90 | 62.90 | 55.30 | 64.70 | 0.61 | 0.65 | 0.51 | 0.65 | 74.76 | 81.24 | 95.70 | 76.41 |
| 5000 | 67.80 | 68.40 | 60.70 | 69.40 | 0.73 | 0.72 | 0.64 | 0.73 | 62.97 | 62.77 | 80.68 | 62.33 |

The results of the pri-awGA (without Hybridization of FLC & LC) and mo-hGA are given in Table 4 and Table 5. While in two problems, mo-hGA got the shortest distance $D1_R$, and also gives better performance than pri-awGA by $R_{NDS}(S_j)$ measure. However, mo-hGA did not effective method combine the number of obtained solutions $|S_j|$. We show the Pareto optimal solutions obtained from pri-awGA and mo-hGA with the test problem of comprises 25 nodes and 49 arcs in Fig. 8.

**Table 4.** Comparison results using the three performance measures at generation *gen*=500

| Test Problems | $\|S_j\|$ | | $R_{NDS}(S_j)$ | | $D1_R$ | |
|---|---|---|---|---|---|---|
| (# of nodes/ # of arcs) | pri-awGA | mo-hGA | pri-awGA | mo-hGA | pri-awGA | mo-hGA |
| 25/49 | 61 | 60 | 0.300 | 0.524 | 4.916 | 1.909 |
| 25/56 | 62 | 56 | 0.258 | 0.428 | 3.946 | 1.829 |

**Table 5.** Comparison results using the three performance measures at generation *gen*=1000

| Test Problems | $\|S_j\|$ | | $R_{NDS}(S_j)$ | | $D1_R$ | |
|---|---|---|---|---|---|---|
| (# of nodes/ # of arcs) | pri-awGA | mo-hGA | pri-awGA | mo-hGA | pri-awGA | mo-hGA |
| 25/49 | 60 | 58 | 0.379 | 0.483 | 1.945 | 0.727 |
| 25/56 | 67 | 60 | 0.516 | 0.597 | 1.647 | 1.207 |

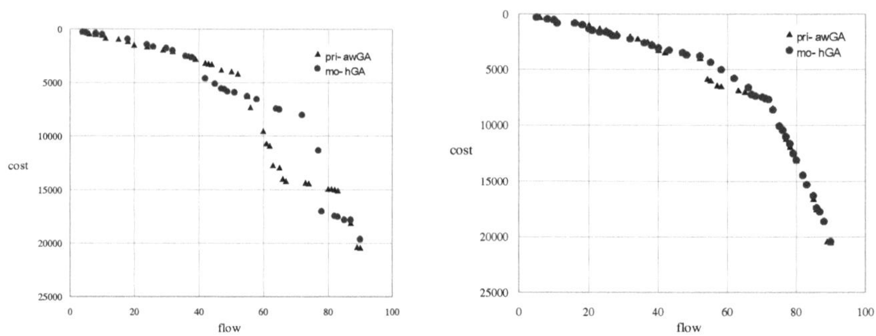


(a) The results is obtained at generation *gen*=50; (b) The results is obtained at generation *gen*=100.

**Fig. 8.** Pareto optimal solutions obtained from pro-awGA and mo-hGA with the test problem of comprises 25 nodes and 49 arcs

# 5 Conclusions

In this paper, we considered the bicriteria network design (bND) problem with the two conflicting objectives of minimizing total cost and maximizing total flow. To solve this bND problem, Special difficulties arises from: (1) a solution of the bND problem is presented by various numbers of paths, (2) a path contains various numbers of nodes and the maximal number is n-1 for an n node network, and (3) a random sequence of edges usually does not correspond to a path. For solving the problem, (1) we proposed a new chromosome representation based on priority-based encoding method. In addition, we gave a special decoding method. We can obtain a solution with variable number of paths of the bND problem. (2) Considering the characteristic of priority-based encoding method, we proposed a new crossover operator called as Weight Mapping Crossover (WMX), and adopted insertion mutation operator and hybrid approach by fuzzy logic control (FLC) and local search (LS). (3) For ensure the population diversity in multiobjective GA, we proposed an interactive adaptive-weight fitness assignment approach. Their elements represent that weights are adjusted adaptively based on the current generation to obtain search pressure toward the positive ideal point, and different from Pareto ranking-based fitness assignment, the sorting process is unnecessary, thus it is effective for considering the computation time. The effectiveness and efficiency of proposed GA approach were investigated with recent related researches.

**Acknowledgments.** This work is partly supported by the Ministry of Education, Science and Culture, the Japanese Government: Grant-in-Aid for Scientific Research (No. 17510138, No.19700071, No. 20500143).

# References

1. Gen, M., Cheng, R.W.: Genetic Algorithm and Engineering Design. Wiley, New York (1997)
2. Marathe, M.V., Ravi, R., Sundaram, R., Ravi, S.S., Rosenkrantz, D.J., Hunt, H.B.: Bicriteria network design problems. J. of Algorithms 28(1), 142–171 (1998)
3. Yuan, D.: A bicriteria optimization approach for robust OSPF routing. In: Proc. of IPOM, pp. 91–98 (2003)

4. Yang, H., Maier, M., Reisslein, M., Carlyle, W.M.: A genetic algorithm-based methodology for optimizing multiservice convergence in a metro WDM network. J. of Lightwave Technology 21(5), 1114–1133 (2003)
5. Raghavan, S., Ball, M.O., Trichur, V.S.: Bicriteria product design optimization (2001), `http://techreports.isr.umd.edu/ARCHIVE/`
6. Zhou, G., Min, H., Gen, M.: A genetic algorithm approach to the bi-criteria allocation of customers to warehouses. Inter. J. of Production Economics 86, 35–45 (2003)
7. Garey, M., Johnson, D.: Computers and Intractability: A Guide to the Theory of NP-Completeness. W.H. Freeman, New York (1979)
8. Chankong, V., Haimes, Y.Y.: Multiobjective Decision Making Theory and Methodology. North-Holland, New York (1983)
9. Warburto, A.: Approximation of Pareto optima in multiple-objective, shortest path problems. Operations Research 35(1), 70–79 (1987)
10. Hassin, R.: Approximation schemes for the restricted shortest path problem. Math. of Operations Research 17(1), 36–42 (1992)
11. Ravi, R.: Rapid rumor ramification: approximating the minimum broadcast time. In: Proc. of 35th Annual IEEE Foundations of Computer Science, pp. 202–213 (1994)
12. Ganley, J.L., Golin, M.J., Salowe, J.S.: The multi-weighted spanning tree problem. In: Li, M., Du, D.-Z. (eds.) COCOON 1995. LNCS, vol. 959, pp. 141–150. Springer, Heidelberg (1995)
13. Munakata, T., Hashier, D.J.: A genetic algorithm applied to the maximum flow problem. In: Proc. of 5th Inter. Conf. on Genetic Algorithms, pp. 488–493 (1993)
14. Hwang, C., Yoon, K.: Multiple Attribute Decision Making: Methods and Applications. Springer, Berlin (1981)
15. Gen, M., Cheng, R., Lin, L.: Network Models and Optimization: Multiobjective Genetic Algorithm Approach. Springer, Heidelberg (2008)
16. Cheng, R., Gen, M.: Evolution program for resource constrained project scheduling problem. In: Proc. IEEE Int. Conf. on Evolutionary Computation, pp. 736–741 (1994)
17. Gen, M., Cheng, R., Wang, D.: Genetic algorithms for solving shortest path problems. In: Proc. IEEE Int. Conf. on Evolutionary Computation, pp. 401–406 (1997)
18. Lin, L., Gen, M., Cheng, R.: Priority-based Genetic Algorithm for Shortest Path Routing Problem in OSPF. In: Proc. 3rd Inter. Conf. on Information and Management Sciences, Dunhuang, China, pp. 411–418 (2004)
19. Whitley, D., Gordan, V., Mathias, K.: Lamarckian evolution, the Baldwin effect & function optimization. In: Davidor, Y., Männer, R., Schwefel, H.-P. (eds.) PPSN 1994. LNCS, vol. 866, pp. 6–15. Springer, Heidelberg (1994)
20. Gen, M., Cheng, R.: Genetic Algorithm and Engineering Optimization. Wiley, New York (2000)
21. Ishibuchi, H., Murata, T.: A multiobjective genetic local search algorithm and its application to flowshop scheduling. IEEE Transactions on Systems, Man, and Cybernetics 28(3), 392–403 (1998)
22. Zitzler, E., Thiele, L.: Multiobjective evolutionary algorithms: a comparative case study and the strength Pareto approach. IEEE Transactions on Evolutionary Computation 3(4), 257–271 (1999)
23. Deb, K., Pratap, A., Agarwal, S., Meyarivan, T.: A fast and elitist multiobjective genetic algorithm: NSGA-II. IEEE Transactions on Evolutionary Computation 6(2), 182–197 (2002)
24. Munakata, T., Hashier, D.J.: A genetic algorithm applied to the maximum flow problem. In: Proc. of 5th Inter. Conf. on Genetic Algorithms, pp. 488–493 (1993)

# Hybrid Genetic Algorithm for Designing Logistics Network, VRP and AGV Problems

Mitsuo Gen[1], Lin Lin[1], and Jung-Bok Jo[2]

[1] Graduate School of Information, Production and Systems, Waseda University
gen@waseda.jp, linlin@aoni.waseda.jp
[2] Division of Computer and Information Engineering, Dongseo University
Phone: +81-90-9565-2964
jobok@gdsu.dongseo.ac.kr

**Abstract.** The use of hybrid genetic algorithm (Hybrid GA) in the networks design has been growing the last decades due to the fact that many practical networks design problems are NP hard. This paper examines recent developments in the field of evolutionary optimization for network design. We combine the various hybrid genetic algorithms to a wide range of practical network problems such as a logistics network model, VRP (Vehicle Routing Problem), and AGV (Automated Guided Vehicles) dispatching problem. It is covered as follows: first, we apply the hybrid priority-based GA for solving fixed-charge Transportation Problem (fcTP), which the proposed approach is more effective in larger size than benchmark test problems. Second, we give the several resent GA approach for solving Multistage Logistic Network Problem. Third, we introduce Vehicle Routing Problem (VRP) and variants of VRP. We apply the priGA for solving Multi-depot vehicle routing problem with time windows (mdVRP-tw). Lastly, we apply a priority-based based GA to solve an automated guided vehicles (AGV) dispatching problem in Flexible Manufacturing System (FMS).

**Keywords:** Hybrid Genetic Algorithms, Network Design, Fixed-charge Transportation Problem (fcTP), Multistage Logistic Network Problem, VRP (Vehicle Routing Problem), AGV (Automated Guided Vehicles) dispatching.

## 1 Introduction

Network model is being increasingly important in the fields such as engineering, computer science, operations research, transportation, telecommunication, decision support systems, manufacturing, and scheduling. Because any system or structure which may be considered abstractly as a set of elements, certain pairs of which are related in a special way, has a representation as a network. Networks provide a useful way for modeling real world problems and are extensively used in many different types of systems: communications, hydraulic, mechanical, electronic, and logistics. Many real world applications impose on more complex issues, such as, complex structure, complex constraints, and multiple objectives to be handled simultaneously and make the problem intractable to traditional approaches [1] [2].

Recent advances in Evolutionary Computation (EC) have made it possible to solve such practical network design problems. As one of EC methods, Genetic Algorithm (GA) is one of the most powerful and broadly applicable stochastic search and

M. Gen et al.: Intelligent and Evolutionary Systems, SCI 187, pp. 123–139.
springerlink.com 

optimization techniques based on principles from evolution theory. In the past few years, the genetic algorithms community has turned much of its attention toward the optimization of network design problems. However, for many GA applications, especially for the problems from network design problems, the simple GA approach is difficult to be applied directly. Thus, how to design efficient algorithms suitable for complex cases (NP-hard) of network design problems by GA technique is a key issue of this research work.

The general form of GA was described by Goldberg [3]. GA is one of the stochastic search algorithms based on the mechanism of natural selection and natural genetics. GA, differing from conventional algorithms, starts with an initial set of random solutions called population $P(t)$. Each individual in the population is called an individuals (or chromosome), representing a potential solution to the problem. The individuals evolve through successive iterations, called generations.

During each generation, the individuals are evaluated by using some measures of fitness. To create the next generation, new chromosomes called offspring $C(t)$ are formed by either merging two individuals from current generation using crossover operator and/or modifying an individual using mutation operator. A new generation is formed by the selection of good individuals according to their fitness values. After several generations, the algorithm converges to the best individual, which hopefully represents the optimal solution or near-optimal solution for the problem. Fig. 3.1 shows a general structure of GA. In general, a GA has five basic components, as summarized by Michalewicz [4][5]:

(1) A genetic representation of solutions to the problem.
(2) A way to create an initial set of potential solutions.
(3) An evaluation function rating solutions in terms of their fitness.
(4) Genetic operators that alter the genetic composition of offspring (crossover, mutation, selection, etc.).
(5) Values for the parameters of genetic algorithms (population size, probabilities of genetic operators, etc.).

Cheng and Gen proposed priority-based Genetic Algorithm (priGA) firstly for solving Resource-constrained Project Scheduling Problem (rcPSP) [6]. Gen *et al.* [7] and Lin *et al.* [8] also adopted this method for solving SPP problem. The priority-based encoding method is an indirect approach. As it is known, a gene in a chromosome is characterized by two factors: locus, *i.e.*, the position of gene located within the structure of chromosome, and allele, *i.e.*, the value the gene takes. In this encoding method, the position of a gene is used to represent task ID and its value is used to represent the priority of the task for constructing assignment sequence among candidates. In this paper, we developed priGA with different decoding procedure to considering the characteristic of these logistics optimization problems.

In this paper, we consider a wide range of network design problems, from the basic transportation models to multistage logistic networks and automated guided vehicles dispatching. The rest of the paper is organized as follows. First, we examine effectiveness of priGA for solving fixed-charge Transportation Problem (see Sect. 2).

Two-stage Logistic Network Problems (tsLNP) (see Sect.3), Vehicle Routing Problem (VRP) (see Sect.4) and Automated Guided Vehicles (AGVs) Dispatching (see Sect.5). Finally, the Section 6 draws the conclusion of this paper.

## 2 Basic Transportation Models

The transportation problem (TP) was formulated and proposed by Hitchcock [9]. Although this problem might seen almost too simple to have much applicability, the TP is very important in our real life applications.

### 2.1 Basic Version of TP

The basic version of the TP is linear, single objective, balanced, and planar problem. Because the problem possesses a special structure in its constraints, an efficient optimization algorithm has been proposed for it, which is the variation of the simplex method adapted to the particular structure [10].

***Nonlinear side constrained Transportation Problem:*** The transportation problem with nonlinear side constraints (nsc-TP) has many applications in our real world [11]. *Exclusionary side constrained Transportation Problem:* In this model, the TP is extended to satisfy the additional constraint in which the simultaneous shipment from some pairs of source centers is prohibited. With this additional side constraint, the problem becomes enormously more difficult, yet the relevance for the real world applications also increases significantly. *Fixed-charge Transportation Problem:* Linear transportation problem is well-known as the simplest model of distribution problem [15]. But fcTP is much more difficult to solve, due to the presence of fixed-charges, which cause discontinuities in the objective function. fcTP has a wide variety of classic applications that have been documented in the scheduling and facility location literature. Two of the most common of these arise (1) in making warehouse or plant location decisions, where there is a charge for opening the facility, and (2) in transportation problems, where there are fixed-charges for transporting goods between demand and supply points [12]. In the fcTP, two types of costs are considered simultaneously when the best course of action is selection: (1) variable costs proportional to the activity level; (2) fixed costs.

***Indices:*** $i$ is index of plant ($i$=1,2,...,$m$), $j$ is index of warehouse ($j$=1,2,...$n$)

***Parameters:*** $a_i$ means number of units available at plant $i$, $b_j$ means number of units demanded at warehouse $j$, $c_{ij}$ means the cost of shipping one unit from plants $i$ to warehouse $j$, $d_{ij}$ means fixed cost associated with route $(i, j)$

***Decision variables:*** $x_{ij}$ means the unknown quantity to be transported on route $(i, j)$, $f_{ij}(\boldsymbol{x})$ total transportation cost for shipping per unit from plant $i$ to warehouse $j$ in which $f_{ij}(\boldsymbol{x})=c_{ij}x_{ij}$ will be a cost function if it is linear.

The usual objective function is to minimize the total variable cost and fixed costs from the allocation. It is one of the combinatorial problems involving constraints. This fcTP with $m$ plants and $n$ warehouses can be formulated as follows:

$$\min\ f(x)=\sum_{i=1}^{m}\sum_{j=1}^{n}\left[f_{ij}(x)+d_{ij}g_{ij}(x)\right] \tag{2.1}$$

$$\text{s.t.}\ \sum_{j=1}^{n}x_{ij}\le a_i,\qquad i=1,2,\dots,m \tag{2.2}$$

$$\sum_{i=1}^{m}x_{ij}\ge b_j,\qquad j=1,2,\dots,n \tag{2.3}$$

$$x_{ij}\ge 0,\qquad \forall i,j \tag{2.4}$$

$$\text{with}\ g_{ij}(x)=\begin{cases}1, & \text{if } x_{ij}>0\\ 0, & \text{otherwise}\end{cases}$$

While equation (2.2) and (2.3) ensure the satisfaction of the plant's capacity and warehouse's demand. Equation (2.4) enforces the non-negativity restriction on the decision variable.

## 2.2 Hybrid GA with Local Search

***Representation:*** For solving the fcTP, a chromosome $v_k(l)$ ($l$=1, 2,..., $L$, $k$=1,2,..., *popSize*, where *popSize* is total number of chromosomes in each generation) consists of priorities of plants and warehouses to obtain transportation tree, and its length is equal to total number of plants ($m$) and warehouses ($n$). Only one arc is added to tree for selecting a plant (warehouse) with the highest priority and connecting it to a warehouse (plant) which considers minimum unit cost.

Fig. 1 shows the representation of fcTP with 3 plants and 7 warehouses. From first to third gene represents 3 plants and the others represent 7 warehouses.

| node ID ($l$) | 1 | 2 | 3 | 4 | 5 | 6 | 7 | 8 | 9 | 10 |
|---|---|---|---|---|---|---|---|---|---|---|
| priority $v_k(l)$ | 5 | 1 | 7 | 10 | 4 | 9 | 2 | 6 | 3 | 8 |

**Fig. 1.** Sample representation by priority-based encoding

***Genetic Operators:*** Crossover and Mutation, we use genetic operators as follows: Partial-Mapped Crossover (PMX) and the Swap mutation are used. PMX uses a special repairing procedure to resolve the illegitimacy caused by the simple two-point crossover. Thus the essentials of PMX are a simple two-point crossover plus a repairing procedure. Swap mutation is used, which simply selects two positions at random and swaps their contents [13].

***Evaluation and selection:*** Evaluation function used for the GA is based on total transportation cost for shipping per unit and the fixed cost from plant $i$ to warehouse $j$ in this problem. The evaluation function is related to the objective function. Therefore, the evaluation function using total cost is defined as follows:

$$eval(v_k)=1/f(x)=1\Big/\sum_{i=1}^{m}\sum_{j=1}^{n}\left[f_{ij}(x)+d_{ij}g_{ij}(x)\right]$$

For the selection methods, we use elitist method that enforces the best chromosomes into the next generation. Because in elitism ensure that at least one copy of the best individual in the population is always passed onto the next generation, the convergence is guaranteed.

***Local Search Techniques:*** The idea of combining genetic algorithms (GAs) with local search (LS) techniques for solving optimization problems has been investigated extensively during the past decade, and various methods of hybridization have been proposed.

Since hybrid approach can combine the merits of GA with those of LS technique, the hybrid approach with GA is less likely to be trapped in a local optimum than LS technique alone.

GAs are used to global exploration among the population of GA, while LS techniques perform local exploitation around the convergence area of GA. Because of the complementary properties between GAs and LS techniques, the hybrid approach often outperforms either the former or the latter alone.

One of the most common forms of hybrid GA is to incorporate a LS technique into a conventional GA loop. With the hybrid GA, the LS technique is applied to each newly generated offspring to move it to a local optimum before injecting it into the population of GA [13]. In this study, we adopt LS technique which is applied to each newly generation of GA, select the best individual, and use insertion mutation until the offspring which the fitness is better than the best individual in offsprings $v_c$ is generated and insert it into the population [16].

## 2.3 Numerical Experiments and Conclusions

We tested 4 problems taken from fcTP benchmark problems [17]. A comparison between our proposed algorithm and the best known results is described in this section. All experiments were realized using JAVA language under Pentium IV PC with 2.6 GHz CPU and 1GB RAM. Each simulation was run 30 times. GA parameter settings were taken as follows: Population size: *popSize* =100; Maximum generation: *maxGen* =1000; Crossover probability, $p_C$ = 0.70; Mutation probability, $p_M$ = 0.50; Terminating condition, *T*=200 generations with the best solution not improved.

**Table 1.** The computational results of each test problem

| Test problems | Best | s-GA | | | ls-hGA | | |
|---|---|---|---|---|---|---|---|
| (# of plants× # of DC) | Known[12] | Worst | Average | Best | Worst | Average | Best |
| ran 10× 10 (a) | 1499 | 1626 | 1613.2 | 1589 | 1626 | 1610.9 | 1589 |
| ran 10× 10 (b) | 3073 | 3264 | 3177.6 | **3073** | 3311 | 3157.8 | **3073** |
| ran 10× 10 (c) | 13007 | 13090 | 13027.0 | **13020** | 13090 | 13034.0 | **13020** |
| ran 13× 13 | 3252 | 3522 | 3498.5 | 3424 | 3538 | 3482.3 | **3286** |

Table 1 shows the computational results of simple GA (sGA) and hybrid GA with Local Search (ls-hGA) to each test problem. By using ls-hGA, we can get same solutions and better solutions comparing s-GA in all test problems. The proposed ls-hGA can find the same solution in ran 10×10 (b), and near-best solution in ran 10×10 (c), ran 13×13. As explained above, we can find best solution and near-best solution by proposed ls-hGA approach. For more realistic problem, we generated 3 problems randomly larger size than fcTP benchmark problems.

**Table 2.** The computational results of three large-size problems

| Test problems | s-GA | | | ls-hGA | | |
|---|---|---|---|---|---|---|
| (# of plants× # of DC) | Worst | Average | Best | Worst | Average | Best |
| ran 20 × 50 | 10305 | 10156.1 | 9921 | 10183 | 9759.1 | **9306** |
| ran 30 × 70 | 13955 | 13685.1 | 13212 | 12911 | 12592.8 | **12123** |
| ran 40 × 100 | 25333 | 25053.2 | 24198 | 24081 | 23221.7 | **22478** |

We simulated three problems 30 times ran 20×50, ran 30×70, ran 40×100, respectively. GA parameter settings were as same as described above. The computational results are show in Table 2. Comparing s-GA with ls-hGA, we can get better solutions in all large-size problems. The proposed approach is effective to solve not only benchmark problems but large-size problems.

## 3 Multistage Logistic Networks

Multistage logistic network design is to provide an optimal platform for efficient and effective logistic systems. This problem and its different versions have been studied in literature [18] [20-23].

### 3.1 Two-Stage Logistic Networks

The efficiency of the logistic system is influenced by many factors; one of them is to decide the number of DCs, and find the good location to be opened, in such a way that the customer demand can be satisfied at minimum DCs' opening cost and minimum shipping cost. In this paper, we consider an extension of two-stage logistic network problem (tsLNP). The problem aims to determine the transportation network to satisfy the customer demand at minimum cost subject to the plant and DCs capacity and also the maximum number of DCs to be opened. Most companies have only limited resources to open and operate DCs. So, limiting the number of DCs that can be located is important when a manager has limited available capital. For this reason, the maximum number of DCs to be opened is considered as constraint in this study.

The tsLNP considered in the study aims to determine the distribution network to satisfy the customer demand at minimum cost subject to the plant and DCs capacity and also the minimum number of DCs to be opened. We assumed that the customer locations and their demand were known in advance. The numbers of potential DC locations as well as their maximum capacities were also known. The mathematical model of the problem is:

$$\min \quad Z = \sum_{i=1}^{I}\sum_{j=1}^{J} t_{ij}x_{ij} + \sum_{j=1}^{J}\sum_{k=1}^{K} c_{jk}y_{jk} + \sum_{j=1}^{J} g_j z_j \tag{3.1}$$

$$\text{s. t.} \quad \sum_{j=1}^{J} x_{ij} \le a_i, \qquad \forall i \tag{3.2}$$

$$\sum_{k=1}^{K} y_{jk} \le b_j z_j, \qquad \forall j \tag{3.3}$$

$$\sum_{j=1}^{J} z_j \le W, \tag{3.4}$$

$$\sum_{j=1}^{J} y_{jk} \ge d_k, \qquad \forall k \tag{3.5}$$

$$\sum_{i=1}^{I}\sum_{j=1}^{J} x_{ij} = \sum_{j=1}^{J}\sum_{k=1}^{K} y_{jk} \tag{3.6}$$

$$x_{ij}, y_{jk} \ge 0, \qquad \forall i, j, k \tag{3.7}$$

$$z_j = \{0,1\} \qquad \forall j \tag{3.8}$$

where, $I$: number of plants ($i$ = 1,2,…,$I$), $J$ : number of distribution centers ($j$ = 1,2,…,$J$), $K$ : number of customers ($k$=1,2,…,$K$), $a_i$ : capacity of plant $i$, $b_j$ : capacity of distribution center $j$, $d_k$ : demand of customer $k$, $t_{ij}$ : unit cost of transportation from plant $i$ to distribution center $j$, $c_{jk}$ : unit cost of transportation from distribution center $j$ to customer $k$, $g_j$ : fixed cost for operating distribution center $j$, $W$ : an upper limit on total number of DCs that can be opened, $x_{ij}$ : the amount of shipment from plant $i$ to distribution center $j$, $y_{jk}$ : the amount of shipment from distribution center $j$ to customer $k$, $z_j$ : 0-1 variable that takes on the value 1 if DC $j$ is opened.

While constraints (3.2) and (3.3) ensure that the plant-capacity constraints and the distribution center-capacity constraints, respectively, constraint (3.4) satisfies the opened DCs do not exceed their upper limit. This constraint is very important when a manager has limited available capital. Constraint (3.5) ensure that all demand of customers are satisfied by opened DCs; Constraints (3.6) and (3.7) enforce the non-negativity restriction on the decision variables and the binary nature of the decision variables used in this model. Without loss of generality, we assume that this model satisfies the balanced condition, since the unbalanced problem can be changed balanced one by introducing dummy suppliers or dummy customers.

## 3.2 Priority-Based Genetic Algorithm

***Representation:*** Michalewicz [4] was the first researcher, who used GA for solving linear and non-linear transportation/distribution problems. In their approach, matrix-based representation had been used. When $m$ and $n$ are the number of sources and depots, respectively, the dimension of matrix is $m \times n$. Although representation is very simple, there is need to special crossover and mutation operators for obtaining feasible solutions.

The use of spanning tree GA (st-GA) for solving some network problems was introduced by [5][13]. They employed Prüfer number in order to represent a candidate solution to the problems and developed feasibility criteria for Prüfer number to be decoded into a spanning tree. They noted that the use of Prüfer number is very suitable for encoding a spanning tree, especially in some research fields, such as transportation problems, minimum spanning tree problems, and so on.

In this study, to escape from these repair mechanisms in the search process of GA, we propose a new encoding method based on priority-based encoding developed. For the problem, a chromosome consists of priorities of sources and depots to obtain transportation tree and its length is equal to total number of sources ($m$) and depots ($n$), i.e. $m+n$. The transportation tree corresponding with a given chromosome is generated by sequential arc appending between sources and depots. At each step, only one arc is added to tree selecting a source (depot) with the highest priority and connecting it to a depot (source) considering minimum cost. Fig. 2 represents a transportation tree with 4 sources and 5 depots, its cost matrix and priority based encoding.

***Genetic operators:*** In this study, we propose a new crossover operator called as weight mapping crossover (WMX) and investigate the effects of four different crossover operators on the performance of GA. WMX can be viewed as an extension of one-point crossover for permutation encoding. As in one-point crossover, after determining a random cut-point, the offspring are generated by using left segment of the cut-point and caring out remapping on the right segment of own parent. In the remapping process, after obtaining an increasing order of digits on the right segments of parents and mapping digits on the ordered parts, new right segment of the first offspring is obtained using original sequence of right segment of the second parent and its mapped digits on the first parent. When obtaining new right segment of second parent, original sequence of right segment of the first parent and its mapped digits on the second parent are used [23]. We also investigate the effects of two different mutation operators on the performance of GA. Insert and swap mutations are used for this purpose.

$$c_{jk} = \begin{array}{c} \\ 1 \\ 2 \\ 3 \\ 4 \end{array} \begin{array}{c} \begin{array}{ccccc} 1 & 2 & 3 & 4 & 5 \end{array} \\ \begin{bmatrix} 15 & 18 & 16 & 19 & 21 \\ 17 & 20 & 12 & 18 & 15 \\ 25 & 23 & 15 & 21 & 19 \\ 20 & 16 & 13 & 14 & 18 \end{bmatrix} \end{array}$$

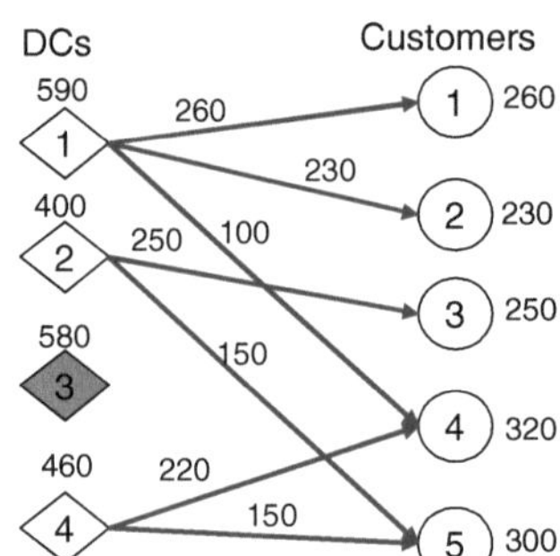


| node ID $l$ : | 1 | 2 | 3 | 4 | 1 | 2 | 3 | 4 | 5 |
|---|---|---|---|---|---|---|---|---|---|
| priority $v(l)$ : | 7 | 5 | 2 | 4 | 6 | 3 | 9 | 1 | 8 |

**Fig. 2.** A sample of transportation tree and its encoding

### 3.3 Numerical Examples

To investigate the effectiveness of the developed GA with new encoding method (priGA), we used spanning tree-based GA (st-GA) using Prüfer number proposed by [24]. Seven different test problems were considered.

Table 3 gives computational results for the st-GA and priGA based on Prüfer number encoding and priority-based encoding methods, respectively, on seven test problems. In st-GA, one-cutpoint crossover and insertion mutation operators were used as genetic operators and its rates were taken as 0.5. Each test problem is run by 10 times using GA approaches. To make comparison between st-GA and priGA according to solution quality and computational burden, we consider again best, average and worst costs and also ACT. In addition, each test problem is divided into three numerical experiments to investigate the effects of population size and number of generations on the performance of st-GA and priGA. When we compare columns of the best cost of the st-GA and priGA, it is possible to see that priGA developed in this study reaches optimum solutions for the first four test problems, while st-GA finds optimum solution for only the first problem. In addition, average percent deviation from optimum solution on st-GA changes between 2.31% and 30% except to the first problem. For big size problems, i.e. last three problems, the best costs of priGA are always smaller than found with st-GA.

**Table 3.** Computational Results for st-GA and priGA

| Problem | Parameters | | st-GA | | | | priGA | | | |
|---|---|---|---|---|---|---|---|---|---|---|
| | *popSize* | *maxGen* | Best | Average | Worst | ACT* | Best | Average | Worst | ACT* |
| 1 | 10 | 300 | 1089 | 1175.4 | 1339 | 0.07 | 1089 | 1089.0 | 1089 | 0.12 |
| | 15 | 500 | 1089 | 1091.8 | 1099 | 0.16 | 1089 | 1089.0 | 1089 | 0.23 |
| | 20 | 1000 | 1089 | 1089.0 | 1089 | 0.35 | 1089 | 1089.0 | 1089 | 0.57 |
| 2 | 20 | 1000 | 2341 | 2402.5 | 2455 | 0.48 | 2283 | 2283.2 | 2285 | 0.78 |
| | 30 | 1500 | 2291 | 2375.2 | 2426 | 1.06 | 2283 | 2283.0 | 2283 | 1.76 |
| | 50 | 2000 | 2303 | 2335.8 | 2373 | 2.42 | 2283 | 2283.0 | 2283 | 4.10 |
| 3 | 30 | 1500 | 2781 | 2874.4 | 2942 | 1.25 | 2527 | 2527.0 | 2527 | 2.04 |
| | 50 | 2500 | 2719 | 2787.1 | 2874 | 3.43 | 2527 | 2527.0 | 2527 | 5.91 |
| | 100 | 4000 | 2623 | 2742.2 | 2796 | 11.85 | 2527 | 2527.0 | 2527 | 21.32 |
| 4 | 75 | 2000 | 3680 | 3873.8 | 4030 | 7.78 | 2886 | 2891.2 | 2899 | 12.99 |
| | 100 | 3000 | 3643 | 3780.4 | 3954 | 15.93 | 2886 | 2892.6 | 2899 | 26.85 |
| | 150 | 5000 | 3582 | 3712.5 | 3841 | 41.41 | 2886 | 2890.0 | 2893 | 71.76 |
| 5 | 75 | 2000 | 5738 | 5949.1 | 6115 | 18.29 | 2971 | 2985.3 | 3000 | 29.07 |
| | 100 | 3000 | 5676 | 5786.1 | 5889 | 36.88 | 2967 | 2980.6 | 2994 | 59.13 |
| | 150 | 5000 | 5461 | 5669.4 | 5835 | 94.33 | 2952 | 2973.2 | 2989 | 153.02 |
| 6 | 100 | 2000 | 7393 | 7705.6 | 8067 | 36.27 | 2975 | 2999.0 | 3025 | 56.32 |
| | 150 | 3000 | 7415 | 7563.8 | 7756 | 76.23 | 2963 | 2994.3 | 3005 | 130.29 |
| | 200 | 5000 | 7068 | 7428.5 | 7578 | 188.37 | 2962 | 2984.9 | 3000 | 295.28 |
| 7 | 100 | 2000 | 10474 | 11083.1 | 11306 | 177.03 | 3192 | 3204.2 | 3224 | 241.74 |
| | 150 | 3000 | 10715 | 10954.7 | 11146 | 395.52 | 3148 | 3184.3 | 3207 | 548.30 |
| | 200 | 5000 | 10716 | 10889.4 | 11023 | 875.03 | 3136 | 3179.6 | 3202 | 1213.65 |

ACT*: Average computation time as second.

## 4 Vehicle Routing Problem Models

Vehicle routing problem (VRP) is a generic name given to a whole class of problems in which a set of routes for a fleet of vehicles based at one or several depots must be determined for a number of geographically dispersed cities or customers. The objective of the VRP is to deliver a set of customers with known demands on minimum-cost vehicle routes with minimum number of vehicles originating and terminating at a depot.

VRP is a well known integer programming problem which falls into the category of NP-hard problems, meaning that the computational effort required solving this problem increase exponentially with the problem size. For such problems it is often desirable to obtain approximate solutions, so they can be found fast enough and are sufficiently accurate for the purpose. Usually this task is accomplished by using various heuristic methods, which rely on some insight into the problem nature [25].

***Capacitated VRP (cVRP):*** cVRP is a VRP in which a fixed fleet of delivery vehicles of uniform capacity must service known customer demands for a single commodity at minimum transit cost.

***VRP with time windows (VRP-tw):*** The time window constant is denoted by a predefined time interval, given an earliest arrival time and latest time. Each customer also imposes a service time to the route, taking consideration of the service time of goods.

***VRP with Pick-up and Delivery (VRP-pd):*** VRP-pd is a VRP in which the possibility that customers return some commodities is contemplated. So in VRP-pd it's needed to take into account that the goods that customers return to the deliver vehicle must fit into it.

***VRP with simultaneous Pick-up and Delivery (VRP-sPD):*** The problem dealing with a single depot distribution/collection system servicing a set of customers by means of a homogeneous fleet of vehicles. Each customer requires two types of service, a pickup and a delivery. The critical feature of the problem is that both activities have to be carried out simultaneously by the same vehicle (each customer is visited exactly once). Products to be delivered are loaded at the depot and products picked up are transported back to the depot. The objective is to find the set of routes servicing all the customers at the minimum cost.

***VRP with Backhauls (VRP-b):*** VRP-b is a VRP in which customers can demand or return some commodities. So in VRP-pd it's needed to take into account that the goods that customers return to the deliver vehicle must fit into it. The critical assumption in that all deliveries must be made on each route before any pickups can be made. This arises from the fact that the vehicles are rear-loaded, and rearrangement of the loads on the tracks at the delivery points is not deemed economical or feasible. The quantities to be delivered and picked-up are fixed and known in advance.

***Multiple Depot VRP (mdVRP):*** A company may have several depots from which it can serve its customers. The mdVRP can be solved in two stages: first, customers must assigned to depots; then routes must be built that link customers assigned to the same depot.

***Split Delivery VRP (sdVRP):*** sdVRP is a relaxation of the VRP wherein it is allowed that the same customer can be served by different vehicles if it reduces overall costs. This relaxation is very important if the sizes of the customer orders are as big as the capacity of a vehicle.

## 4.1 Problem Description (mdVRP-tw)

To solve multi-depot VRP-tw (mdVRP-tw), when the number of customers is usually much larger than that of DC, we can adopt cluster approach first, and then route ones.

mdVRP-tw become more complex as it involves servicing customers with time windows using multiple vehicles that vary in number with respect to the problem. Therefore, mdVRP-tw should be designed as follows: (1) All distances are represented by Euclidean distance. (2) Each customer is serviced by one of depots. (3) Each route starts a depot and then returns the depot. (4) Each customer can be visited only once by a vehicle. (5) The vehicle capacity of each route is equal. (6) Total customer demand for each route does not exceed the vehicle capacity. (7) Each customer is associated with a time window period for its service time. (8) Each vehicle has maximum travel time.

The objective for solving mdVRP-tw is to determine depot and vehicle routing system in order to achieve the minimal cost without violating the DC capacity and time window constraints. mdVRP-tw is an NP-hard problem due to an NP-hard of VRP-tw. The mdVRP-tw is to determine the set of vehicle routing that can satisfy the customer demand within its time-window constraints, thus, we divided it into two phases. First phase is to cluster customers and then vehicle routing phase is considered.

## 4.2 Genetic Algorithms

**Clustering customers (Phase 1):** The aim of this phase is to determine the assignment of customers to each DC so that the total distance is minimized. Parallel assignment: We adopted parallel assignment for clustering customers. The name parallel is due to the fact that the urgency for each customer is calculated considering all depots at the same time [26].

**Vehicle routing (Phase 2):** The aim of this phase is to develop the vehicle routing from DC satisfying time window constraint.

**Genetic representation:** In this step, we propose GA with priority-based encoding method to escape the repair mechanisms in the search process of GA.

| node ID ($l$) | 1 | 2 | 3 | 4 | 5 | 6 | 7 | 8 | 9 | 10 | 11 | 12 |
|---|---|---|---|---|---|---|---|---|---|---|---|---|
| priority $v(l)$ | 4 | 12 | 1 | 11 | 2 | 9 | 3 | 6 | 7 | 5 | 8 | 10 |

**Fig. 3.** Sample representation by priority-based encoding

All the customers are sorted in increasing order of earliest arrival time. We use the sorted customer number by node ID in a chromosome. The sample representation by priority-based encoding is represented in Fig. 3.

At each step, only one customer is added to set selected by the highest priority and find the next customer considering minimum distance. We consider the sequence of route, first assigned customer form DC is r, the next is $u$, $u$+1, and so on. In time window constraints, we have to consider start time at customer $j$ $t_j^S$, which is the time of starting time to the next customer. Finish time at customer $j$ $t_j^F$ means the time of finishing the service at customer $j$. We also consider not only the customer which is selected by the highest priority but also left and right gene from it. In the encoding procedure, we take the new priority divided by the ID No from original priority. By using this method, we can assign more customers in a route. The sample representation by new priority-based encoding is represented in Fig. 4.

| node ID ($l$) | 1 | 2 | 3 | 4 | 5 | 6 | 7 | 8 | 9 | 10 | 11 | 12 |
|---|---|---|---|---|---|---|---|---|---|---|---|---|
| priority $v(l)$ | 4.00 | 6.00 | 0.33 | 2.75 | 0.40 | 1.50 | 0.42 | 0.75 | 0.77 | 0.50 | 0.72 | 0.83 |

**Fig. 4.** The sample representation by new priority-based encoding

**Crossover and Mutation:** We use genetic operators as follows: Order Crossover (OX) and the Swap mutation are used. It can be viewed as a kind of PMX that uses a different repair procedure. Swap mutation is used, which simply selects two positions at random and swaps their contents.

### 4.3 Numerical Experiments

To prove the efficiency of the proposed GA approaches, we tested several problems comparing the result of two approaches. In this study, we generated six test problems and each problem consists of small size (2 DCs / 60 customers) and large size (3DCs /100 customers). The geographical data are randomly generated in each problem. Maximum load of vehicles is 150 in all test problems. We also consider three factors for more realistic vehicle routing problem: (1) Capacities of DCs. (2) A mix of short scheduling and a long scheduling in a problem. (3) Different service time for customers.

All of problems are represented in Appendix. We test 6 problems by using proposed GA and represents the customer routes and total distances. All experiments were realized using C language under Pentium IV PC with 2.7 GHz CPU and 1GB RAM. GA parameter settings were taken as follows: Population size: *popSize* =100; Maximum generation: *maxGen* =1500; Crossover probability, $p_C$ = 0.70; Mutation probability, $p_M$ = 0.50; Terminating condition, $T$=200 generations with the best solution not improved. Table 4 represents the fleet of vehicles and total distance of each test problem.

**Table 4.** Computational results of each test problems

| Test No. | # of DCs / # of customers | Proposed GA-1 | |
|---|---|---|---|
| | | NV | TD |
| 1-1 | 2 / 60 | 12 | 982.334 |
| 1-2 | 3 /100 | 20 | 1771.903 |
| 2-1 | 2 / 60 | 12 | 826.374 |
| 2-2 | 3 /100 | 17 | 1472.461 |
| 3-1 | 2 / 60 | 13 | 878.753 |
| 3-2 | 3 /100 | 18 | 1489.279 |

## 5 Automated Guided Vehicles Dispatching

Automated material handling has been called the key to integrated manufacturing. An integrated system is useless without a fully integrated, automated material handling system. In the manufacturing environment, there are many automated material handling possibilities. Currently, automated guided vehicles systems (AGV Systems), which include automated guided vehicles (AGVs), are the state–of–the–art, and are often used to facilitate automatic storage and retrieval systems (AS/RS) [28].

In this study, we focus on the simultaneous scheduling and routing of AGVs in a flexible manufacturing system (FMS). A FMS environment requires a flexible and adaptable material handling system. AGVs provide such a system. An AGV is a material handling equipment that travels on a network of guide paths. The FMS is composed of various cells, also called working stations (or machine), each with a specific operation such as milling, washing, or assembly. Each cell is connected to the guide path network by a pickup/delivery (P/D) point where pallets are transferred from/to the AGVs. Pallets of products are moved between the cells by the AGVs.

### 5.1 Network Modeling for AGV Dispatching

In this paper, the problem is to dispatch AGVs for transports the product between different machines in a FMS. At first stage, we model the problem by using network structure. Assumptions considered in this paper are as follows: *For FMS scheduling*: (1) In a FMS, $n$ jobs are to be scheduled on $m$ machines. (2) The $i$-th job has $n_i$ operations that have to be processed. (3) Each machine processes only one operation at a time. (4) The set-up time for the operations is sequence-independent and is included in the processing time. *For AGV dispatching*: (1) Each machine is connected to the guide path network by a pick-up/delivery (P/D) station where pallets are transferred from/to the AGVs. (2) The guide path is composed of aisle segments on which the vehicles are assumed to travel at a constant speed. (3) As many vehicles travel on the guide path simultaneously, collisions be avoided by hardware, not be considered in this paper. Subject to the constraints that: *For FMS scheduling*: (1) The operation sequence for each job is prescribed; (2) Each machine can process only one operation at a time; (3) Each AGV can transport only one kind of products at a time. *For AGV dispatching*: (1) AGVs only carry one kind of products at same time. (2) The vehicles just can travel forward, not backward.

The objective function is minimizing the time required to complete all jobs (*i.e.* makespan): $t_{\text{MS}}$. The problem can be formulated as follows:

$$\min \quad t_{\text{MS}} = \max_i \left\{ t_{i,n_i}^S + t_{M_{i,n_i},0} \right\} \tag{5.1}$$

$$\text{s.t.} \quad c_{ij}^S - c_{i,j-1}^S \geq p_{i,j-1} + t_{ij}, \ \forall i, j = 2, \ldots, n_i \tag{5.2}$$

$$\left(c_{ij}^S - c_{i'j'}^S - p_{i'j'} + \Gamma\left|M_{ij} - M_{i'j'}\right| \geq 0\right) \vee \left(c_{i'j'}^S - c_{ij}^S - p_{ij} + \Gamma\left|M_{ij} - M_{i'j'}\right| \geq 0\right), \ \forall (i,j),(i',j') \tag{5.3}$$

$$\left(t_{ij}^S - t_{i'j'}^S - t_{i'j'} + \Gamma\left|x_{ij} - x_{i'j'}\right| \geq 0\right) \vee \left(t_{i'j'}^S - t_{ij}^S - t_{ij} + \Gamma\left|x_{ij} - x_{i'j'}\right| \geq 0\right), \ \forall (i,j),(i',j') \tag{5.4}$$

$$\left(t_{i,n_i}^S - t_{i'j'}^S - t_{i'j'} + \Gamma\left|x_{ij} - x_{i'j'}\right| \geq 0\right) \vee \left(t_{i'j'}^S - t_{i,n_i}^S - t_i + \Gamma\left|x_{ij} - x_{i'j'}\right| \geq 0\right), \ \forall (i,n_i),(i',j') \tag{5.5}$$

$$c_{ij}^S \geq t_{i,j+1}^S - p_{ij} \tag{5.6}$$

$$x_{ij} \geq 0, \ \forall i, j \tag{5.7}$$

$$t_{ij}^S \geq 0, \ \forall i, j \tag{5.8}$$

where Γ is a very large number, and $t_i$ is the transition time for pickup point of machine $M_{i,ni}$ to delivery point of Loading/ Unloading. Inequality (5.2) describes the operation precedence constraints. In inequities (5.3), (5.4) and (5.5), since one or the other constraint must hold, it is called disjunctive constraint. It represents the operation un-overlapping constraint (Inequality 5.3) and the AGV non-overlapping constraint (Inequality 5.4, 5.5).

## 5.2 Priority-Based GA

We firstly give a priority-based encoding method that is an indirect approach: encode some guiding information for constructing a sequence of all tasks. As it is known, a gene in a chromosome is characterized by two factors: locus, *i.e.*, the position of gene located within the structure of chromosome, and allele, *i.e.*, the value the gene takes. In this encoding method, the position of a gene is used to represent task ID and its value is used to represent the priority of the task for constructing a sequence among candidates. A feasible sequence can be uniquely determined from this encoding with considering operation precedence constrain. An example of generated chromosome and its decoded path is shown as following:

| Task ID : | 1 | 2 | 3 | 4 | 5 | 6 | 7 | 8 | 9 |
|---|---|---|---|---|---|---|---|---|---|
| Priority : | 1 | 5 | 7 | 2 | 6 | 8 | 3 | 9 | 4 |

$$T_{11} \rightarrow T_{12} \rightarrow T_{13} \rightarrow T_{14} \rightarrow T_{21} \rightarrow T_{22} \rightarrow T_{31} \rightarrow T_{32} \rightarrow T_{33}$$

At the beginning, we try to find a task for the position next to source node s. Task $T_{11}$, $T_{21}$ and $T_{31}$ (Task ID: 1, 2 and 3) are eligible for the position, which can be easily fixed according to adjacent relation among tasks. The priorities of them are 1, 5 and 7, respectively. The node 1 has the highest priority and is put into the task sequence. The possible tasks next to task $T_{11}$, is task $T_{12}$ (Task ID: 4), and unselected task $T_{21}$ and $T_{31}$ (Task ID: 2 and 3). Because node 4 has the largest priority value, it is put into the task sequence. Then we form the set of tasks available for next position and select the one with the highest priority among them. Repeat these steps until all of tasks be selected,

$$T_{11} \rightarrow T_{12} \rightarrow T_{13} \rightarrow T_{14} \rightarrow T_{21} \rightarrow T_{22} \rightarrow T_{31} \rightarrow T_{32} \rightarrow T_{33}$$

After generated task sequence, we secondly separate tasks to several groups for assigning different AGVs. First, separate tasks with a separate point in which the task is the final transport of job $i$ form pickup point of operation $O_{i,\,ni}$ to delivery point of Loading/Unloading. Afterward, unite the task groups which finished time of a group is faster than the starting time of another group. The particular is introduced in next subsection. An example of grouping is shown as following:

$$\begin{aligned}
&\text{AGV } 1: T_{11} \rightarrow T_{12} \rightarrow T_{13} \rightarrow T_{14}\\
&\text{AGV } 2: T_{21} \rightarrow T_{22}\\
&\text{AGV } 3: T_{31} \rightarrow T_{32} \rightarrow T_{33}
\end{aligned}$$

## 5.3 Case Study

For evaluating the efficiency of the AGV dispatching algorithm suggested in a case study, a simulation program was developed by using Java on Pentium 4 processor (3.2-GHz clock). The problem was used by [19] [27]. GA parameter settings were taken as follows: population size, *popSize* =20; maximum generation, *maxGen*=1000; crossover probability, $p_C$=0.70; mutation probability, $p_M$ =0.50; immigration rate, $\mu = 0.15$.

In a case study of FMS, 10 jobs are to be scheduled on 5 machines. The maximum number process for the operations is 4. Table 5 gives the assigned machine numbers and process time. And Table 6 gives the transition time among pickup points and delivery points.

We can draw a network depend on the precedence constraints among tasks $\{T_{ij}\}$ of case study. The best result of case study is shown as follows and final time required to complete all jobs (*i.e.* makespan) is 574 and 4 AGVs are used. Fig. 5 shows the result on Gantt chart.

**Table 5.** Job Requirements of Example

| $O_{ij}$ | $M_{ij}$ | | | | $p_{ij}$ | | | |
|---|---|---|---|---|---|---|---|---|
| $J_i$ \ $P_j$ | $P_1$ | $P_2$ | $P_3$ | $P_4$ | $P_1$ | $P_2$ | $P_3$ | $P_4$ |
| $J_1$ | 1 | 2 | 1 | - | 80 | 120 | 60 | - |
| $J_2$ | 2 | 1 | - | - | 100 | 60 | - | - |
| $J_3$ | 5 | 3 | 3 | - | 70 | 100 | 70 | - |
| $J_4$ | 5 | 3 | 2 | 2 | 70 | 100 | 100 | 40 |
| $J_5$ | 4 | 2 | - | - | 90 | 40 | - | - |
| $J_6$ | 4 | 4 | 1 | 2 | 90 | 70 | 60 | 40 |
| $J_7$ | 1 | 3 | - | - | 80 | 70 | - | - |
| $J_8$ | 5 | 4 | 5 | 4 | 70 | 70 | 70 | 80 |
| $J_9$ | 5 | 4 | 1 | - | 70 | 70 | 60 | - |
| $J_{10}$ | 5 | 1 | 3 | - | 70 | 60 | 70 | - |

**Table 6.** Transition Time between Pickup Point *u* and Delivery Point *v*

| $t_{uv}$ / $c_{uv}$ | Loading / Unloading | $M_1$ | $M_2$ | $M_3$ | $M_4$ | $M_5$ |
|---|---|---|---|---|---|---|
| Loading / Unloading | 1 / 1 | 1 / 7 | 8 / 13 | 14 / 18 | 16 / 23 | 18 / 20 |
| $M_1$ | 13 / 18 | 3 / 3 | 2 / 9 | 8 / 14 | 10 / 19 | 13 / 18 |
| $M_2$ | 18 / 22 | 22 / 28 | 2 / 2 | 2 / 7 | 4 / 12 | 12 / 18 |
| $M_3$ | 13 / 11 | 17 / 22 | 24 / 29 | 1 / 1 | 1 / 6 | 7 / 11 |
| $M_4$ | 8 / 14 | 12 / 20 | 18 / 26 | 24 / 29 | 3 / 3 | 2 / 10 |
| $M_5$ | 5 / 7 | 9 / 12 | 15 / 18 | 19 / 23 | 23 / 28 | 2 / 2 |

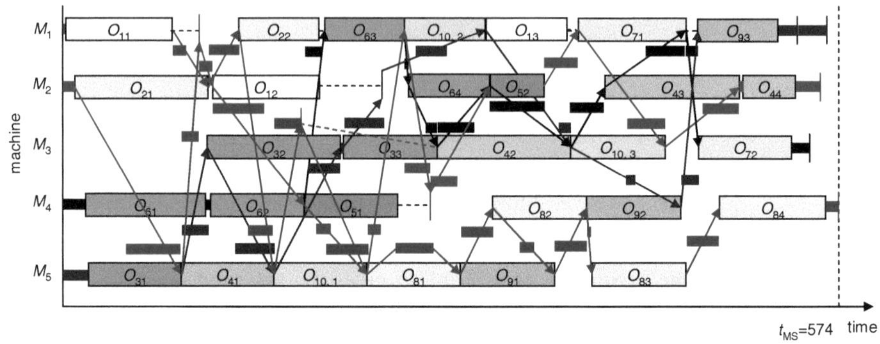


**Fig. 5.** Gantt chart of the schedule of Example with considering AGVs routing based on operations processing

# 6 Conclusions

The use of priority-based Genetic Algorithm (priGA) in the network design has been growing the last decades due to the fact that practical network design problem is often NP hard. This paper examined recent developments in the field of evolutionary optimization for network design problems in various areas. A wide range of problem is covered as follows: first, we applied the hybrid priGA approach for solving fixed-charge transportation problem. We have done several numerical experiments and compared the results with those of the traditional GA. The proposed approach is more effective in larger size than benchmark test problems. Second, we gave the several resent GA approaches for solving multistage logistic network problems. Third, we introduced vehicle routing problem (VRP) and variants of VRP. We apply the priGA for solving multi-depot vehicle routing problem with time windows. Lastly, we modelled an automated guided vehicle (AGV) system by using network structure. This network model of an AGV dispatching has simplexes decision variables with considering most AGV problem's constraints. Furthermore, we applied priGA for solving this problem with minimizing time required to complete all jobs (*i.e.*, makspan).

**Acknowledgments.** This work is partly supported by the Ministry of Education, Science and Culture, the Japanese Government: Grant-in-Aid for Scientific Research (No.17510138, No.19700071, No. 20500143).

# References

1. Gen, M., Cheng, R., Lin, L.: Network Models and Optimization: Multiobjective Genetic Algorithm Approach. Springer, London (2008)
2. Gen, M., Kumar, A., Kim, J.R., et al.: Recent network design techniques using evolutionary algorithms. International Journal of Production Economics 98(2), 251–261 (2005)
3. Goldberg, D.: Genetic Algorithms in Search, Optimization and Machine Learning. Addison-Wesley, Reading (1989)
4. Michalewicz, Z.: Genetic Algorithm + Data Structures = Evolution Programs, revised edn. Springer, New York (1996)
5. Gen, M., Cheng, R.: Genetic Algorithm and Engineering Desig. Wiley, New York (1997)

6. Cheng, R., Gen, M.: Evolution program for resource constrained project scheduling problem. In: Proc. IEEE Int. Conf. on Evolutionary Computation, pp. 736–741 (1994)
7. Gen, M., Cheng, R., Wang, D.: Genetic algorithms for solving shortest path problems. In: Proc. IEEE Int. Conf. on Evolutionary Computation, pp. 401–406 (1997)
8. Lin, L., Gen, M., Cheng, R.: Priority-based Genetic Algorithm for Shortest Path Routing Problem in OSPF. In: Proc. 3rd Inter. Conf. on Information and Management Sciences, pp. 411–418 (2004)
9. Hitchcock, F.: The distribution of a product from several sources to numerous locations. Journal of Mathematical Physics 20, 224–230 (1941)
10. Bazaraa, M., Jarvis, J., Sherali, H.: Linear Programming and Network Flows, 2nd edn. John Wiley & Sons, New York (1993)
11. Cao, B., Uebe, G.: Solving Transportation Problems with Nonlinear Side Constraints with Tabu Search. Computer & Ops. Res. 22(6), 593–603 (1995)
12. Adlakha, V., Kowalski, K.: Simple heuristic algorithm for the solution of small fixed-charge problems. Omega, Int. Journal of Management Science 31, 205–211 (2003)
13. Gen, M., Cheng, R.: Genetic Algorithm and Engineering Optimization. Wiley, New York (2000)
14. Altiparmak, F., Gen, M., Lin, L., Paksoy, T.: A genetic algorithm approach for multi-objective optimization of supply chain networks. Computer & Industrial Engineering 51, 197–216 (2006)
15. Jo, J.B., Li, Y., Gen, M.: Nonlinear fixed-charge transportation problem by spanning tree-based genetic algorithm. Computer & Industrial Engineering 53, 290–298 (2007)
16. Gen, M., Lin, L.: Multiobjective hybrid genetic algorithm for bicriteria network design problem. In: Proc. of Asia Pacific Symposium on Intelligent and Evolutionary Systems, vol. 8, pp. 73–82 (2004)
17. Gamsworld (accessed), `http://www.gamsworld.org`
18. Syarif, A., Yun, Y., Gen, M.: Study on multi-stage logistics chain network: a spanning tree-based genetic algorithm approach. Computers and Industrial Engineering 43, 299–314 (2002)
19. Yang, J.B.: GA-Based Discrete Dynamic Programming Approach for Scheduling in FMS Environment. IEEE Trans. on Sys, Man, and Cyb. -B 31(5), 824–835 (2001)
20. Jayaraman, V., Ross, A.: A simulated annealing methodology to distribution network design and management. European Journal of Operational Research 144, 629–645 (2003)
21. Gen, M., Syarif, A.: Hybrid genetic algorithm for multi-time period production/distribution planning. Computers and Industrial Engineering, Computers & Industrial Engineering 48(4), 799–809 (2005)
22. Gen, M.: Study on Evolutionary Network Design by Multiobjective Hybrid Genetic Algorith. PhD dissertation, 123, Kyoto University (2006)
23. Gen, M., Altiparamk, F., Lin, L.: A Genetic Algorithm for Two-stage Transportation Problem using Priority-based Encoding. OR Spectrum 28(3), 337–354 (2006)
24. Syarif, A., Gen, M.: Solving exclusionary side constrained transportation problem by using a hybrid spanning tree-based genetic algorithm. Journal of Intelligent Manufacturing 14, 389–399 (2003)
25. VRP Web (accessed), `http://neo.lcc.uma.es/radi-eb/WebVRP/`
26. Tansini, L., Urquhart, M., Viera, O.: Comparing assignment algorithms for the Multi-Depot VRP. Jornadas de Informática e Investigación Operativa (1999)
27. Kim, K., Yamazaki, G., Lin, L., Gen, M.: Network-based Hybrid Genetic Algorithm to the Scheduling in FMS environments. J. of Artificial Life and Robotics 8(1), 67–76 (2004)
28. Naso, D., Turchiano, B.: Multicriteria meta-heuristics for AGV dispatching control based on computational intelligence. IEEE Trans. on Sys. Man & Cyb.-B 35(2), 208–226 (2005)

# Multiobjective Genetic Algorithm for Bicriteria Network Design Problems

Lin Lin and Mitsuo Gen

Graduate School of Information, Production and Systems, Waseda University
linlin@aoni.waseda.jp, gen@waseda.jp

**Abstract.** Network design is one of the most important and most frequently encountered classes of optimization problems. However, various network optimization problems typically cannot be solved by a generalized approach. Usually we must design the different algorithm for the different type of network optimization problem depending on the characteristics of the problem. In this paper, we try to investigate with a broad spectrum of multi-criteria network design models, analyze the recent related researches, design and validate new effective multiobjective hybrid genetic algorithms for three kinds of major bicriteria network design models: bicriteria shortest path (bSP) model, bicriteria minimum spanning tree (bST) model and bicriteria network flow (bNF) model. Because of the adaptability, robustness and flexibility of the evolutionary algorithms, proposed approaches are easy applied to many kinds of real applications extended from these major network design models.

**Keywords:** Multiobjective hybrid genetic algorithms, bicriteria shortest path model, bicriteria minimum spanning tree model and bicriteria network flow model.

## 1 Introduction

Network design is one of the most important and most frequently encountered classes of optimization problems [1]. It is a combinatory field in graph theory and combinatorial optimization. A lot of optimization problems in network design arose directly from everyday practice in engineering and management: determining shortest or most reliable paths in traffic or communication networks, maximal or compatible flows, or shortest tours; planning connections in traffic networks; coordinating projects; and solving supply and demand problems. Furthermore, network design is also important for complexity theory, an area in the common intersection of mathematics and theoretical computer science which deals with the analysis of algorithms. However, there is a large class of network optimization problems for which no reasonable fast algorithms have been developed. And many of these network optimization problems arise frequently in applications. Given such a hard network optimization problem, it is often possible to find an efficient algorithm whose solution is approximately optimal. Among such techniques, the genetic algorithm (GA) is one of the most powerful and broadly applicable stochastic search and optimization techniques based on principles from evolution theory.

Network design problems where even one cost measure must be minimized are often NP-hard [2]. However, in practical applications it is often the case that the network to be built is required to multiobjective. In the following, we introduce three core bicriteria network design models. (1) *Bicriteria shortest path* (bSP) model is one

M. Gen et al.: Intelligent and Evolutionary Systems, SCI 187, pp. 141–161.
springerlink.com 

of the basic multi-criteria network design problems. It is desired to find a diameter constrained path between two specified nodes with minimizing two cost functions. Hansen presented the first bSP model [3]. Recently, Skriver and Andersen examined the correlative algorithms for the bSP problems [4]; Azaron presented a new methodology to find the bicriteria shortest path under the steady-state condition [5]. (2) *Bicriteria minimum spanning tree* (bMST) model play a central role within the field of multi-criteria network modes. It is desired to find a subset of arcs which is a tree and connects all the nodes together with minimizing two cost functions. Marathe *et al.* presented a general class of bST model [6], and Balint proposed a non-approximation algorithm to minimize the diameter of a spanning sub-graph subject to the constraint that the total cost of the arcs does not exceed a given budget [7]. (3) *Bicriteria maximum flow* (bMXF) model and bSP model are mutual complementary topics. It is desired to send as much flow as possible between two special nodes without exceeding the capacity of any arc. Lee and Pulat presented algorithm to solve a bNF problem with continuous variables [8]. (4) *Bicriteria network flow* (bNF) model: as we know, the shortest path problem (SPP) considers arc flow costs but not flow capacities; the maximum flow (MXF) problem considers capacities but only the simplest cost structure. SPP and MXF combine all the basic ingredients of network design problems. Bicriteria network flow model is an *integrated bicriteria network design* (bNF) model integrating these nuclear ingredients of SPP and MXF. This bNF model considers the flow costs, flow capacities and multiobjective optimization.

The bicriteria network design models provide useful ways to model real world problems, which are extensively used in many different types of complex systems such as communication networks, manufacturing systems and logistics systems. For example, in a communication network, we want to find a set of links which consider the connecting cost (or delay) and the high throughput (or reliability) for increasing the network performance [9] [10]; as an example in the manufacturing application described in [11], the two criteria under consideration are cost, that we wish to minimize, and manufacturing yield, that we wish to maximize; in a logistics system, the main drive to improve logistics productivity is the enhancement of customer services and asset utilization through a significant reduction in order cycle time (lead time) and logistics costs [12].

Recently, genetic algorithm (GA) and other evolutionary algorithms (EAs) have been successfully applied in a wide variety of network design problems [13]. For example, Ahn and Ramakrishna developed a variable-length chromosomes and a new crossover operator for shortest path routing problem [14], Wu and Ruan (2004) proposed a gene-constrained GA for solving shortest path problem [15], Li *et al.* (2006) proposed a specific GA for *optimum path planning* in *intelligent transportation system* (ITS) [16], Kim *et al.* (2007) proposed a new path selection scheme which uses GA along with the modified roulette wheel selection method for *MultiProtocol label switching* (MPLS) network [17], Hasan *et al.* (2007) proposed a *novel heuristic GA* to solve the *single source shortest path* (ssSP) problem [18], Ji *et al.* developed a *simulation-based GA* to find multi-objective paths with minimizing both expected travel time and travel time variability in ITS [19], Chakraborty *et al.* developed *multiobjective genetic algorithm* (moGA) to find out simultaneously several alternate routes depending on distance, contains minimum number of turns, path passing through mountains [20], Garrozi and Araujo presented a moGA to solve the *multicast routing problem* with maximizing the common links in source-destination routes and minimizing the route sizes [21], and Kleeman *et al.* proposed a *modified nondominated sorting genetic algorithm II* (nsGA II) for the *multicommodity capacitated network design problem* (mcNDP), the multiple objectives including costs, delays, robustness, vulnerability, and reliability [22].

The paper is organized as follows: In Section 2, we give three kinds of major bicriteria network design models: bicriteria shortest path (bSP) model, bicriteria minimum spanning tree (bST) model and bicriteria network flow (bNF) model. In Section 3, we investigate with a broad spectrum of the recent related researches, design new effective multiobjective hybrid genetic algorithms for the bicriteria network design models. In Section 4, we demonstrate effectiveness comparing with different encoding methods, and also demonstrate effectiveness comparing with different multiobjective GAs. This paper give the conclusion follows in Section 5.

## 2 Bicriteria Network Design Models

Let $G=(N, A)$ be a directed network, consisting of a finite set of nodes $N = \{1, 2, \ldots, n\}$ and a set of directed arcs $A = \{(i, j), (k, l), \ldots, (s, t)\}$ joining $m$ pairs of nodes in $N$. Arc $(i, j)$ is said to be incident with nodes $i$ and $j$, and is directed from node $i$ to node $j$. Suppose that each arc $(i, j)$ has assigned to it nonnegative numbers $c_{ij}$, the cost of $(i, j)$ or other parameters of $(i, j)$. Let $x_{ij}$ is a decision variables the link on an arc $(i, j)\in A$.

### 2.1 Bicriteria Shortest Path (bSP) Model

The *shortest path model* is the heart of network design optimization. In this paper, let $d_{ij}$ is transmission delay of arc $(i, j)$, we consider the bSP model of finding minimizing total cost $z_1$ and minimizing delay $z_2$ from a source node $s$ (node 1) to a sink node $t$ (node $n$), the bSPP can be defined by the following assumptions:

A1. The network is directed. We can fulfill this assumption by transforming any undirected network into a directed network.
A2. All transmission delay and all arc costs are nonnegative.
A3. The network does not contain parallel arcs (*i.e.,* two or more arcs with the same tail and head nodes). This assumption is essentially a notational convenience.

The bSP problem is formulated as follows:

$$\textbf{min} \quad z_1 = \sum_{i=1}^{n}\sum_{j=1}^{n} c_{ij}x_{ij} \tag{1}$$

$$\textbf{min} \quad z_2 = \sum_{i=1}^{n}\sum_{j=1}^{n} d_{ij}x_{ij} \tag{2}$$

$$\textbf{s.t.} \quad \sum_{j=1}^{n} x_{ij} - \sum_{k=1}^{n} x_{ki} = \begin{cases} 1 & (i=1) \\ 0 & (i=2,3,\cdots,n-1) \\ -1 & (i=n) \end{cases} \tag{3}$$

$$x_{ij} = 0 \text{ or } 1 \quad (i, j = 1, 2, \cdots, n) \tag{4}$$

where constraint (3), a conservation law is observed at each of the nodes other than $s$ or $t$. That is, what goes out of node $i$, $\sum_{j=1} x_{ij}$ must be equal to what comes in $\sum_{k=1} x_{ki}$.

## 2.2 Bicriteria Minimum Spanning Tree (bMST) Model

The *Minimum Spanning Tree* (MST) problem is one of the best-known network optimization problems which attempt to find a minimum cost tree network that connects all the nodes in the network. The links or edges have associated costs that could be based on their distance, capacity, quality of line, *etc.*

In this paper, we are considering a bicriteria minimum spanning tree (bMST) model. The bST is to find a set of links with the two conflicting objectives of minimizing communication cost $z_1$ and minimizing the transfer delay $z_2$ and the constraint of network capacity $w_{ij}$ of each edge $(i, j) \in E$ is met.

$$\min \quad z_1(\boldsymbol{x}) = \sum_{i=1}^{n} \sum_{j=1}^{n} c_{ij} x_{ij} \tag{5}$$

$$\min \quad z_2(\boldsymbol{x}) = \sum_{i=1}^{n} \sum_{j=1}^{n} d_{ij} x_{ij} \tag{6}$$

$$\text{s. t.} \quad \sum_{i=1}^{n} \sum_{j=1}^{n} x_{ij} = n - 1 \tag{7}$$

$$\sum_{i=1}^{n} \sum_{j=1}^{n} x_{ij} \le |S| - 1 \quad \text{for any set } S \text{ of nodes} \tag{8}$$

$$\sum_{j=1}^{n} w_{ij} x_{ij} \le u_i, \qquad \forall i \tag{9}$$

$$x_{ij} \in \{0,\ 1\}, \quad \forall i, j \tag{10}$$

where, the 0-1 variable $x_{ij}$ indicates whether we select edge $(i, j)$ as part of the chosen spanning tree (note that the second set of constraints with $|S|=2$ implies that each $x_{ij} \le 1$). The constraint (7) is a cardinality constraint implying that we choose exactly $n$-1 edges, and the packing constraint (8) implies that the set of chosen edges contain no cycles (if the chosen solution contained a cycle, and $S$ were the set of nodes on a chosen cycle, the solution would violate this constraint). The constraint (9) guarantees that the total link weight of each node $i$ does not exceed the upper limit $W_i$.

## 2.3 Bicriteria Network Flow (bNF) Model

Suppose that each arc $(i, j)$ has assigned to it nonnegative numbers $c_{ij}$, the cost of $(i, j)$ and $u_{ij}$, the capacity of $(i, j)$. This capacity can be thought of as representing the maximum amount of some commodity that can "flow" through the arc per unit time in a steady-state situation. Such a flow is permitted only in the indicated direction of the arc, *i.e.*, from $i$ to $j$.

Consider the problem of maximizing total flow $z_1$ and minimizing total cost $z_2$ from a source node $s$ (node 1) to a sink node $t$ (node $n$). The additional assumptions are given as following:

A4. The network does not contain a directed path from node $s$ to node $t$ composed only of infinite capacity arcs. Whenever every arc on a directed path $P$ from note $s$ to note $t$ has infinite capacity, we can send an infinite amount of flow along this path, and therefore the maximum flow value is unbounded.

A5. The network does not contain parallel arcs (*i.e.*, two or more arcs with the same tail and head nodes). This assumption is essentially a notational convenience.

The bNF problem is formulated as follows,

$$\max \quad z_1 = f \tag{11}$$

$$\min \quad z_2 = \sum_{i=1}^{n}\sum_{j=1}^{n} c_{ij}x_{ij} \tag{12}$$

$$\text{s. t.} \quad \sum_{k=1}^{n} x_{ij} - \sum_{i=1}^{n} x_{ki} = \begin{cases} f & (i=1) \\ 0 & (i=2,3,\cdots,n-1) \\ -f & (i=n) \end{cases} \tag{13}$$

$$0 \le x_{ij} \le u_{ij}, \qquad \forall (i,j) \tag{14}$$

$$f \ge 0 \tag{15}$$

where constraint (13), a *conservation law* is observed at each of the nodes other than $s$ or $t$. Constraint (14) is flow capacity. We call any set of numbers $\boldsymbol{x}=(x_{ij})$ which satisfy (13) and (14) a feasible flow, or simply a flow, and $f$ is its value.

## 3 Multiobjective Genetic Algorithm

The inherent characteristics of GAs demonstrate why genetic search is possibly well suited to the multiple objective optimization problems. The basic feature of GAs is the multiple directional and global search by maintaining a population of potential solutions from generation to generation. The *population-to-population* approach is hopeful to explore all Pareto solutions.

GAs do not have much mathematical requirements about the problems and can handle any kind of objective functions and constraints. Due to their evolutionary nature, the GAs can search for solutions without regard to the specific inner workings of the problem. Therefore, it is more hope for solving much complex problems than the conventional methods.

### 3.1 Priority-Based Genetic Algorithm

How to encode a solution of the network design problem into a chromosome is a key issue for GAs. In Holland's work, encoding is carried out using binary strings. For many GA applications, especially for the problems from network design problems, the simple approach of GA was difficult to apply directly. During the 10 years, various nonstring encoding techniques have been created for network routing problems [23]. We need to consider these critical issues carefully when designing a new non-binary application string coding so as to build an effective GA chromosome.

Given a new encoding method, it is usually necessary to examine whether we can build an effective genetic search with the encoding. Several principles have been proposed to evaluate an encoding [24]:

Property 1 (*Space*): Chromosomes should not require extravagant amounts of memory.

Property 2 (*Time*): The time complexity of executing evaluation, recombination and mutation on chromosomes should not be a higher order.

Property 3 (*Feasibility*): A chromosome corresponds to a feasible solution.

Property 4 (*Uniqueness*): The mapping from chromosomes to solutions (decoding) may belong to one of the following three cases: 1-to-1 *mapping*, *n*-to-1 *mapping* and 1-to-*n mapping*. The 1-to-1 mapping is the best among three cases and 1-to-*n* mapping is the most undesir one.

Property 5 (*Locality*): A small change in chromosome should imply a small change in its corresponding solution.

Property 6 (*Heritability*): Offspring of simple crossover (*i.e.*, one-cut point crossover) should correspond to solutions which combine the basic feature of their parents.

How to encode a path in a network is also critical for developing a GA application to network design problems, it is not easy to find out a nature representation. Special difficulty arises from (1) a path contains variable number of nodes and the maximal number is *n*-1 for an *n* node network, and (2) a random sequence of edges usually does not correspond to a path.

Recently, to encode a diameter-constrained path into a chromosome, various common encoding techniques have been created. Munemoto *et. al.* proposed a *variablelength encoding* method for network routing problems in a wired or wireless environment [25]. Ahn and Ramakrishna developed this variable-length representation and proposed a new crossover operator for solving the shortest path routing (SPR) problem [14]. The advantage of variable-length encoding is the mapping from any chromosome to solution (decoding) belongs to 1-*to*-1 *mapping* (uniqueness). The disadvantages are: (1) in general, the genetic operators may generate infeasible chromosomes (illegality) that violate the constraints, generating loops in the paths; (2) repairing techniques are usually adopted to convert an illegal chromosome to a legal one.

Inagaki *et al.* proposed a *fixed-length encoding* method for multiple routing problems [26]. The proposed method are sequences of integers and each gene represents the node ID through which it passes. To encode an arc from node *i* to node *j*, put *j* in the *i*-th locus of the chromosome. This process is reiterated from the source node 1 and terminating at the sink node *n*.

The advantages of fixed-length encoding are: (1) any path has a corresponding encoding (completeness); (2) any point in solution space is accessible for genetic search; (3) any permutation of the encoding corresponds to a path (legality) using the special genetic operators. The disadvantages are: (1) in some cases, *n*-to-1 mapping may occur for the encoding; (2) in general, the genetic operators may generate infeasible chromosomes (illegality), and special genetic operator phase is required. Therefore we lose feasibility and heritability.

Cheng and Gen proposed a priority-based encoding method firstly for solving Resource-constrained Project Scheduling Problem (rcPSP) [27]. Gen *et al.* also adopted priority-based encoding for the solving bSPP problem [28]. Recently, Lin and Gen proposed a *priority-based encoding* method [29]. As is known, a gene in a chromosome is characterized by two factors: locus, *i.e.*, the position of the gene located within the structure of chromosome, and allele, *i.e.*, the value the gene takes. In this encoding method, the position of a gene is used to represent node ID and its value is used to represent the priority of the node for constructing a path among candidates. A path can be uniquely determined from this encoding.

The advantages of the priority-based encoding method are: (1) any permutation of the encoding corresponds to a path (feasibility); (2) most existing genetic operators can be easily applied to the encoding; (3) any path has a corresponding encoding (legality); (4) any point in solution space is accessible for genetic search. However, there is a disadvantage as that $n$-to-1 mapping (uniqueness) may occur for the encoding at some case. Considering the characteristic of priority-based chromosome, we proposed a new crossover operator, called weight mapping crossover (WMX) and adopted insertion mutation.

Depending on the properties of encodings, we summarize the performance of the priority-based encoding method and other introduced encoding methods in Table 1.

**Table 1.** Summarizing the performance of encoding methods

| Chromosome Design | | Space | Time | Feasibility | Uniqueness | Locality | Heritability |
|---|---|---|---|---|---|---|---|
| variable length-based GA | Ahn *et al., IEEE Trans. EC*, 2002 | $m$ | O($m$log$m$) | poor | 1-to-1 mapping | worst | worst |
| fixed length-based GA | Inagaki *et al., Proc. of IEEE ISCS* 1999 | $n$ | O($n$log$n$) | worst | $n$-to-1 mapping | worst | worst |
| priority-based GA | Lin & Gen, *IEEE-EC* 2007 | $n$ | O($n$log$n$) | good | $n$-to-1 mapping | good | good |

## 3.2 PrimPred-Based Genetic Algorithm

In GAs literature, whereas several kinds of encoding methods were used to obtain MSTs, most of them cannot effectuality encode or decode between chromosomes and legality spanning trees. Special difficulty arises from (1) a cardinality constraint implying that we choose exactly $n-1$ edges, and (2) implying any set of chosen edges containing no cycles. We need to consider these critical issues carefully when designing an appropriate encoding method so as to build an effective GA. How to encode a spanning tree $T$ in a graph $G$ is critical for developing a GA to network design problem, it is not easy to find out a nature representation. We summarized the several kinds of classification of encoding methods as follows:

1. Characteristic vectors-based encoding
2. Edge-based encoding
3. Node-based encoding

### 3.2.1 Characteristic Vectors-Based Encoding

Davis *et al.* [30] and Piggott and Suraweera [31] have used a *binary-based encoding method* to represent spanning trees in GAs. A binary-based encoding requires space proportional to $m$ and the time complexities of binary-based encoding is $O(m)$. The mapping from chromosomes to solutions (decoding) may be *1-to-1 mapping*. Bean [32] described a *random keys-based encoding method* for encoding ordering and scheduling problems. Schindler *et al.* [33] and Rothlauf *et al.* [34] further investigated network random keys in an evolution strategy framework. In this encoding, a chromosome is a string of real-valued weights, one for each edge. To decode a spanning tree, the edges are sorted by their weights, and Kruskal's algorithm considers the edges are sorted order.

As for binary-based encoding, random keys-based encoding requires space proportional to $m$ and the time complexities is $O(m)$. Whereas all chromosomes represent feasible solutions, the uniqueness of the mapping from chromosomes to solutions may be $n$-to-1 mapping.

### 3.2.2 Edge-Based Encoding

*Edge-based encoding* is an intuitive representation of a tree. A general edge-based encoding requires space proportional to $n$–1 and the time complexities is $O(m)$. The mapping from chromosomes to solutions (decoding) may be *1-to-1 mapping*. In a complete graph, $m = n(n-1)/2$ and the size of the search space is $2^{n(n-1)/2}$. Edge-based encoding and binary-based encoding have very similar performance in theory. Knowles and Corne [35] proposed a method which improves edge-based encoding. The basis of this encoding is a spanning-tree construction algorithm which is *randomized primal method* (RPM), based on the Prim's algorithm. Raidl and Julstrom [36] gave the method depending on an underlying random spanning-tree algorithm. The mapping from chromosomes to solutions (decoding) may be 1-to-1 mapping. In a complete graph, $m = n(n-1)/2$, the size of the search space is $n^{n-1}$. These encoding methods offer efficiency of time complexity, feasibility and uniqueness. However, offspring of simple crossover and mutation should represent infeasible solutions. Several special genetic operator and repair strategies have been successful, but their limitations weaken the encoding heritability.

### 3.2.3 Node-Based Encoding

***Prüfer number-based encoding:*** Cayley [37] proved the following formula: the number of spanning trees in a complete graph of $n$ nodes is equal to $n^{n-2}$. Prüfer [38] presented the simplest proof of Cayley's formula by establishing a *1-to-1 correspondence* between the set of spanning trees and a set of sequences of $n-2$ integers, with each integer between 1 and $n$ inclusive. The sequence of $n$ integers for encoding a tree is known as the Prüfer number.

***Predecessor-based Encoding:*** A more compact representation of spanning trees is the predecessor or determinant encoding, in which an arbitrary node in $G$ is designated the root, and a chromosome lists each other node's predecessor in the path from the node to the root in the represented spanning tree: if pred($i$) is $j$, then node $j$ is adjacent to node $i$ and nearer the root. Thus, a chromosome is string of length $n-1$ over 1, 2, .., $n$, and when such a chromosome decodes a spanning tree, its edges can be made explicit in time that is $O(n\log n)$.

***PrimPred-based Encoding:*** We improved the predecessor-based encoding that adopted Prim's algorithm in chromosome generating procedure. Prim's algorithm implements the greedy-choice strategy for minimum spanning tree. Starting with an empty tree (one with no edges), the algorithm repeatedly adds the lowest-weight edge $(u, v)$ in $G$ such that either $u$ or $v$, but not both, is already connected to the tree. Considering the characteristic of predecessor-based encoding, they proposed a new crossover and mutation operators. These operators offer locality, heritability, and computational efficiency.

Depending on the properties of encodings, we summarize the performance of proposed PrimPred-based encoding method and other introduced encoding methods in Table 2.

**Table 2.** Summary of the performance of encoding methods

| Representation | | Space | Time | Feasibility | Uniqueness | Locality | Heritability |
|---|---|---|---|---|---|---|---|
| Characteristic Vectors-based | binary-based encoding | $m$ | $O(m)$ | worst | 1-to-1 mapping | worst | worst |
| | Random keys-based encoding | $m$ | $O(m)$ | good | $n$-to-1 mapping | worst | worst |
| Edge-based | General edge-based encoding | $n$ | $O(m)$ | worst | 1-to-1 mapping | worst | worst |
| | Heuristic edge-based encoding | $n$ | $O(n)$ | good | 1-to-1 mapping | poor | poor |
| Node-based | Prüfer number-based encoding | $n$ | $O(n\log n)$ | good | 1-to-1 mapping | worst | worst |
| | Predecessor-based Encoding | $n$ | $O(n\log n)$ | poor | 1-to-1 mapping | worst | worst |
| | PrimPred-based Encoding | $n$ | $O(n\log n)$ | good | 1-to-1 mapping | poor | poor |

### 3.3 Interactive Adaptive-Weight Genetic Algorithm

GA is essentially a kind of meta-strategy method. When applying the GA to solve a given problem, it is necessary to refine upon each of the major components of GA, such as *encoding methods, recombination operators, fitness assignment, selection operators, constraints handling,* and so on, in order to obtain a best solution to the given problem. Because the multiobjective optimization problems are the natural extensions of constrained and combinatorial optimization problems, so many useful methods based on GA have been developed during the past two decades. One of special issues in multiobjective optimization problems is the *fitness assignment mechanism.*

Although most fitness assignment mechanisms are just different approaches and suitable for different cases of multiobjective optimization problems, in order to understanding the development of moGA, we classify algorithms according to proposed years of different approaches:

Generation 1 **Vector Evaluation Approach:**
  vector evaluated GA (veGA), Schaffer [39]
Generation 2 **Pareto Ranking + Diversity:**
  multiobjective GA (moGA), Fonseca and Fleming [40]
  non-dominated sorting GA (nsGA), Srinivas and Deb [41]
Generation 3 **Weighted Sum + Elitist Preserve:**
  random weight GA (rwGA), Ishibuchi and Murata [42]
  adaptive weight GA (awGA), Gen and Cheng [23]

strength Pareto EA II (spEA II), Zitzler and Thiele[43]
non-dominated sorting GA II (nsGA II), Deb *et al.* [47]
interactive adaptive-weight GA (i-awGA), Lin and Gen [46]

***Interactive Adaptive-weight Genetic Algorithm:*** We proposed an interactive adaptive-weight genetic algorithm (i-awGA), which is an improved adaptive-weight fitness assignment approach with the consideration of the disadvantages of weighted-sum approach and Pareto ranking-based approach. We combined a penalty term to the fitness value for all of dominated solutions. Firstly, calculate the adaptive weight $w_i = 1/( z_i^{\max} - z_i^{\min})$ for each objective $i$=1, 2,…, $q$ by using awGA. Afterwards, calculate the penalty term $p(\boldsymbol{v}_k)$=0, if $\boldsymbol{v}_k$ is nondominated solution in the nondominated set $P$. Otherwise $p(\boldsymbol{v}_{k'})$=1 for dominated solution $\boldsymbol{v}_{k'}$. Last, calculate the fitness value of each chromosome by combining the method as follows and we adopted roulette wheel selection as supplementary to the i-awGA.

$$eval(v_k) = \sum_{i=1}^{q} w_i (z_i^k - z_i^{\min}) + p(v_k), \quad \forall k \in popSize$$

### 3.4 Overall Procedure

The overall procedure of proposed multiobjective GA for solving bicriteria network design model is outlined as follows.

**procedure:** moGA for bicriteria network design models
**input:** network data ($N, A, C, U$), GA parameters ($popSize$, $maxGen$, $p_M$, $p_C$)
**output:** Pareto optimal solutions $E$
**begin**
  $t \leftarrow 0$;
  initialize $P(t)$ by encoding routine;
  calculate objectives $z_i$ $(P)$, $i$ = 1, …,$q$ by decoding routine;
  create Pareto $E(P)$ by nondominated routine;
  evaluate $eval(P)$ by i-awGA routine;
  **while** (**not** terminating condition) **do**
    create $C(t)$ from $P(t)$ by crossover routine;
    create $C(t)$ from $P(t)$ by mutation routine;
    create $C(t)$ from $P(t)$ by immigration routine;
    calculate objectives $z_i(C)$, $i$ = 1, …,$q$ by decoding routine;
    update Pareto $E(P,C)$ by nondominated routine;
    evaluate $eval(P,C)$ by i-awGA routine;
    select $P(t+1)$ from $P(t)$ and $C(t)$ by roulette wheel selection routine;
    $t \leftarrow t+1$;
  **end**
  **output** Pareto optimal solutions $E(P,C)$
**end**

## 4 Experiments and Discussions

For each algorithm, 50 runs with Java are performed on Pentium 4 processor (3.40-GHz clock), 3.00GA RAM.

### 4.1 Performance Measures

In order to evaluate the results of each test, we are using the performance measures: average of the best solutions (ABS), percent deviation from optimal solution (PD), standard deviation (SD). We also give a statistical analysis by ANOVA, and give examples of Pareto frontier, convergence patterns for the problems.

For evaluate the performance of multiobjective GAs, reference solution set S* of each test problem was found using all algorithms which be used in computational experiments. Each algorithm was applied to each test problem with much longer computation time and larger memory storage than the other computational experiments. Generally, we used the very large parameter specifications in all algorithms for finding the reference solution set of each test problem. We chose only nondominated solutions as reference solutions by 10 runs of the algorithms for each test problem.

a. The number of obtained solutions $|S_j|$.

b. The ratio of nondominated solutions $R_{NDS}(S_j)$: A straightforward performance measure of the solution set $S_j$ with respect to the $J$ solution sets is the ratio of solutions in $S_j$ that are not dominated by any other solutions in $S$. The $R_{NDS}(S_j)$ measure can be written as follows:

$$R_{NDS}(S_j) = \frac{\left|S_j - \{x \in S_j \mid \exists r \in S^*: r \prec x\}\right|}{|S_j|}$$

c. The distance $D1_R$ measure can be written as follows:

$$D1_R = \frac{1}{|S^*|} \sum_{r \in S^*} \min\{d_{rx} \mid x \in S_j\}$$

where $S^*$ is a reference solution set for evaluation the solution set $S_j$; $d_{xr}$ is the distance between a current solution $x$ and a reference solution $r$.

$$d_{rx} = \sqrt{(f_1(r) - f_1(x))^2 + (f_2(r) - f_2(x))^2}$$

### 4.2 Experiments for bSP Model

In the first experiment, we demonstrate effectiveness comparing with different genetic representations. The 12 test problems (in Table 3) are combined. In order to evaluate the results of each test, we use the single objective by minimizing total cost, and combine average of the best solutions (ABS). In addition, we demonstrate the difference among the quality of solutions obtained by various GA parameter settings and an auto-tuning strategy proposed by Lin and Gen [46]. There are 3 kinds of different GA parameter settings:

Para 1: $popSize$=10, $p_C$=0.3, $p_M$=0.7, $p_I$ = 0.30
Para 2: $popSize$=10, $p_C$=0.5, $p_M$=0.5, $p_I$ = 0.30
Para 3: $popSize$=10, $p_C$=0.7, $p_M$=0.3, $p_I$ = 0.30

In addition, two different stopping criteria are employed. One of them is the number of maximum generations, $maxGen$ =1000. Another stopping criteria is $T$=200. That is, if the best solution is not improved until successive 200 generations, the algorithm will be stopped.

**Table 3.** Network characteristics # of nodes $n$, # of arcs $m$, cost $c$ and delay $d$ for the networks

| ID | $n$ | $m$ | cost $c$ | delay $d$ |
|---|---|---|---|---|
| 1 | 100 | 955 | runif($m$, 10, 50) | runif($m$, 5, 100) |
| 2 | | 959 | rnorm($m$, 50, 10) | rnorm($m$, 50, 5) |
| 3 | | 990 | rlnorm($m$, 1, 1) | rlnorm($m$, 2, 1) |
| 4 | | 999 | 10*rexp($m$, 0.5) | 10*rexp($m$, 2) |
| 5 | 200 | 2040 | runif($m$, 10, 50) | runif($m$, 5, 100) |
| 6 | | 1971 | rnorm($m$, 50, 10) | rnorm($m$, 50, 5) |
| 7 | | 2080 | rlnorm($m$, 1, 1) | rlnorm($m$, 2, 1) |
| 8 | | 1960 | 10*rexp($m$, 0.5) | 10*rexp($m$, 2) |
| 9 | 500 | 4858 | runif($m$, 10, 50) | runif($m$, 5, 100) |
| 10 | | 4978 | rnorm($m$, 50, 10) | rnorm($m$, 50, 5) |
| 11 | | 4847 | rlnorm($m$, 1, 1) | rlnorm($m$, 2, 1) |
| 12 | | 4868 | 10*rexp($m$, 0.5) | 10*rexp($m$, 2) |

Uniform Distribution: runif($m$, min, max).
Normal Distribution: rnorm($m$, mean, sd).
Lognormal Distribution: rlnorm($m$, meanlog, sdlog).
Exponential Distribution: rexp($m$, rate).

**Table 4.** The ABS of 50 Runs by Different GA Parameter Settings with Different Genetic Representations

| ID | optimal | ahnGA | | | priGA | | | |
|---|---|---|---|---|---|---|---|---|
| | | para1 | para2 | para3 | para1 | para2 | para3 | auto-tuning |
| 1 | 47.93 | 47.93 | 47.93 | 47.93 | 47.93 | 47.93 | 47.93 | 47.93 |
| 2 | 210.77 | 232.38 | 234.36 | 244.64 | 224.82 | 224.91 | 228.72 | 224.09 |
| 3 | 1.75 | 2.69 | 2.71 | 2.83 | 2.68 | 2.73 | 2.79 | 2.64 |
| 4 | 17.53 | 37.60 | 39.43 | 47.26 | 36.10 | 35.30 | 34.08 | 34.60 |
| 5 | 54.93 | 60.77 | 62.26 | 65.35 | 57.26 | 57.42 | 58.50 | 56.87 |
| 6 | 234.45 | 276.72 | 288.71 | 295.77 | 269.23 | 268.52 | 273.16 | 270.66 |
| 7 | 1.83 | 2.40 | 2.66 | 3.31 | 2.01 | 2.27 | 2.32 | 1.98 |
| 8 | 22.29 | 47.29 | 49.58 | 57.04 | 41.68 | 45.89 | 44.17 | 41.90 |
| 9 | 70.97 | - | - | - | 72.29 | 75.74 | 77.26 | 70.97 |
| 10 | 218.78 | - | - | - | 276.56 | 276.15 | 284.85 | 272.10 |
| 11 | 3.82 | - | - | - | 5.85 | 6.91 | 6.41 | 5.78 |
| 12 | 20.63 | - | - | - | 60.14 | 57.52 | 61.53 | 52.18 |

"-" means out of memory error.

Table 4 shows the ABS of 50 runs by different GA parameter settings with different genetic representations respectively. As depicted in Table 4, most results of ABS of 50 runs by priGA with auto-tuning operator proposed are better than each of the other combinations, except to the test 4, test 6 and test 8.

In the second experimental study, we demonstrate the performance comparisons of multiobjective GAs for solving bSP problems by different fitness assignment approaches, there are spEA, nsGA II, rwGA and i-awGA. In each GA approach, priority-based encoding was used, and WMX crossover, insertion mutation and auto-tuning operators were used as genetic operators.

As depicted in Table 5, most results of ABS of 50 runs by i-awGA are better than each of the other fitness assignment approach. In addition, we do not say the efficiency of the approach, only depend on the performance measure $|S_j|$ or $R_{NDS}(S_j)$. We can have worst results when compared to another run with a low $R_{NDS}(S_j)$. Therefore we show the proposed i-awGA outperform another approach with the efficiency both of the performance measure $|S_j|$ or $R_{NDS}(S_j)$. In Table 5, the values of $|S_j|$ are given as rational numbers, though the value of $|S_j|$ was defined as the integer number. Because we give an average of $|S_j|$ with 50 runs for comparing the different approaches. Furthermore, the values of $|S_j|$ increases (or decreases) depended on the characteristic of different testing data.

**Table 5.** The ABS of 50 Runs by Different Fitness Assignments

| ID | $\|S_j\|$ | | | | $R_{NDS}(S_j)$ | | | | $D1_R(S_j)$ | | | |
|---|---|---|---|---|---|---|---|---|---|---|---|---|
| | spEA | nsGA | rwGA | i-awGA | spEA | nsGA | rwGA | i-awGA | spEA | nsGA | rwGA | i-awGA |
| 1 | 1.64 | 1.70 | 1.64 | 1.84 | 1.00 | 1.00 | 1.00 | 1.00 | 0.00 | 0.00 | 0.00 | 0.00 |
| 2 | 5.00 | 5.08 | 4.98 | 5.64 | 0.18 | 0.16 | 0.22 | 0.38 | 0.18 | 0.23 | 0.17 | 0.10 |
| 3 | 3.30 | 3.04 | 3.22 | 3.48 | 0.91 | 0.93 | 0.92 | 0.91 | 0.00 | 0.00 | 0.00 | 0.00 |
| 4 | 7.36 | 7.40 | 7.12 | 7.46 | 0.04 | 0.02 | 0.04 | 0.04 | 0.06 | 0.06 | 0.05 | 0.05 |
| 5 | 3.26 | 3.22 | 3.12 | 3.46 | 1.00 | 1.00 | 1.00 | 1.00 | 0.00 | 0.00 | 0.00 | 0.00 |
| 6 | 1.74 | 2.40 | 2.20 | 1.54 | 0.28 | 0.14 | 0.18 | 0.30 | 0.17 | 0.24 | 0.22 | 0.15 |
| 7 | 4.16 | 3.96 | 3.66 | 3.70 | 0.52 | 0.59 | 0.66 | 0.68 | 0.40 | 0.42 | 0.40 | 0.05 |
| 8 | 5.90 | 4.80 | 5.30 | 5.16 | 0.05 | 0.13 | 0.07 | 0.10 | 1.10 | 0.89 | 0.96 | 0.86 |
| 9 | 1.16 | 1.24 | 1.28 | 1.36 | 0.99 | 0.96 | 0.91 | 0.99 | 0.00 | 0.01 | 0.01 | 0.00 |
| 10 | 2.60 | 2.42 | 2.62 | 2.30 | 0.11 | 0.18 | 0.16 | 0.33 | 1.17 | 0.76 | 0.99 | 0.59 |
| 11 | 2.86 | 2.90 | 2.70 | 3.22 | 0.31 | 0.30 | 0.30 | 0.43 | 0.01 | 0.01 | 0.01 | 0.00 |
| 12 | 5.82 | 6.02 | 6.14 | 6.20 | 0.03 | 0.03 | 0.04 | 0.05 | 0.19 | 0.19 | 0.20 | 0.19 |

In Tables 6 and 7, we use ANOVA analysis depended on the $|S_j|$ and $R_{NDS}(S_j)$ in 50 times with test problem 11 to analyze the difference among the quality of solutions obtained by various 4 kinds of different fitness assignment approaches. Analysis of variance (ANOVA) is a collection of statistical models, and their associated procedures, in which the observed variance is partitioned into components due to different explanatory variables. In this experiment, the explanatory variables are the $|S_j|$ and $R_{NDS}(S_j)$ in 50 times by different approaches. If the value of *mean difference* is greater than the reference value *LSD* (Least Significant Difference), that means compared approaches are statistically difference. As shown in Tables 6 and 7, at the significant level of $\alpha = 0.05$, the $F$=3.56 and 3.12 is greater than the reference value of ($F$=2.84), respectively. The difference between our i-awGA and each of the other approaches (spEA, nsGA II or rwGA) is greater than the LSD=0.31 and 0.10, respectively. We can say our i-awGA indeed statistically better than the other approaches.

**Table 6.** ANOVA Analysis with $|S_j|$ in Test Problem 11

| | spEA | nsGA II | rwGA | i-awGA |
|---|---|---|---|---|
| # of data | 50 | 50 | 50 | 50 |
| Mean | 2.86 | 2.90 | 2.70 | 3.22 |
| SD | 0.92 | 0.83 | 0.78 | 0.64 |
| Variance | 0.84 | 0.69 | 0.61 | 0.41 |
| Sum of squares | 42.02 | 34.50 | 30.50 | 20.58 |
| Factors | Sum of squares | Freedom degree | Mean square | F |
| Between groups | 7.12 | 3 | 2.37 | 3.65 |
| Within-groups | 127.60 | 196 | 0.65 | |
| Total | 134.72 | 199 | | |
| F ($\alpha = 0.05$) | 2.68 | | | |
| t ($\alpha = 0.05$) | 1.98 | | | |
| LSD | 0.31 | | | |
| Mean Difference with i-awGA | 0.36 | 0.32 | 0.52 | |

**Table 7.** ANOVA Analysis with $R_{NDS}(S_j)$ in Test Problem 11

| | spEA | nsGA II | rwGA | i-awGA |
|---|---|---|---|---|
| # of data | 50 | 50 | 50 | 50 |
| Mean | 0.31 | 0.30 | 0.30 | 0.43 |
| SD | 0.27 | 0.22 | 0.26 | 0.23 |
| Variance | 0.07 | 0.05 | 0.07 | 0.05 |
| Sum of squares | 3.62 | 2.43 | 3.33 | 2.62 |
| Factors | Sum of squares | Freedom degree | Mean square | F |
| Between groups | 0.57 | 3 | 0.19 | 3.12 |
| Within-groups | 12.01 | 196 | 0.06 | |
| Total | 12.58 | 199 | | |
| F ($\alpha = 0.05$) | 2.68 | | | |
| t ($\alpha = 0.05$) | 1.98 | | | |
| LSD | 0.10 | | | |
| Mean Difference with i-awGA | 0.11 | 0.13 | 0.13 | |

### 4.3 Experiments for bST Model

In this section, our PrimPred-based GA is compared with Zhou and Gen [13] and Raidl and Julstrom [36] for solving several *large scale minimum spanning tree* (MST) problems. For examining the effectiveness of different encoding methods, PrimPred-based GA, Zhou and Gen's Prüfer number-based encoding method and Raidl and Julstrom's edge-based encoding method are applied to six test problems [45]. Prüfer number-based encoding with one-cut point crossover and swap mutation is combined, and edge-based encoding using two kinds of mutation operators is combined which is included in [44], and for initializing the chromosomes based on the edge set, Raidl and Julstrom's PrimRST (Prim random spanning tree) is combined. Each algorithm was run 20 times using different initial seeds for each test problems. And Prim's algorithm has been used to obtain optimum solutions for the problems. The GA parameter is setting as follows:

Population size: *popSize* =10;
Crossover probability: $p_C$ =0.30, 0.50 or 0.70;
Mutation probability: $p_M$ =0.30, 0.50 or 0.70;
Maximum generation: *maxGen* =1000;

**Table 8.** Performance comparisons with different GA approaches

| Test Problem | Optimal Solutions | $n$ | $m$ | $p_C$ | $p_M$ | Prüfer Num-based | | Edge-based 1 | | Edge-based 2 | | PrimPred-based | |
|---|---|---|---|---|---|---|---|---|---|---|---|---|---|
| | | | | | | avg. | CPU time | avg. | CPU time | avg. | CPU time | avg. | CPU time |
| 1 | 470 | 40 | 780 | 0.30 | 0.30 | 1622.20 | 72.20 | 1491.80 | 1075.20 | 493.60 | 1081.40 | 470.00 | 1100.20 |
| | | | | 0.50 | 0.50 | 1624.40 | 87.60 | 1355.80 | 2184.40 | 505.80 | 2175.00 | 470.00 | 2256.40 |
| | | | | 0.70 | 0.70 | 1652.60 | 134.80 | 1255.20 | 3287.40 | 497.60 | 3281.40 | 470.00 | 3316.00 |
| 2 | 450 | 40 | 780 | 0.30 | 0.30 | 1536.60 | 74.80 | 1458.20 | 1118.60 | 471.60 | 1093.80 | 450.00 | 1106.20 |
| | | | | 0.50 | 0.50 | 1549.20 | 78.20 | 1311.40 | 2190.80 | 480.20 | 2175.00 | 450.00 | 2200.20 |
| | | | | 0.70 | 0.70 | 1564.40 | 122.00 | 1184.40 | 3287.60 | 466.40 | 3262.40 | 450.00 | 3275.00 |
| 3 | 820 | 80 | 3160 | 0.30 | 0.30 | 3880.40 | 150.00 | 3760.20 | 5037.80 | 923.20 | 5059.60 | 820.00 | 5072.00 |
| | | | | 0.50 | 0.50 | 3830.00 | 184.40 | 3692.00 | 10381.20 | 871.00 | 10494.20 | 820.00 | 10440.60 |
| | | | | 0.70 | 0.70 | 3858.20 | 231.20 | 3483.80 | 16034.80 | 899.20 | 15871.80 | 820.00 | 15984.60 |
| 4 | 802 | 80 | 3160 | 0.30 | 0.30 | 3900.60 | 131.40 | 3853.00 | 5125.00 | 894.60 | 4934.20 | 802.00 | 5071.80 |
| | | | | 0.50 | 0.50 | 3849.60 | 206.20 | 3515.20 | 10325.20 | 863.00 | 10268.80 | 802.00 | 10365.60 |
| | | | | 0.70 | 0.70 | 3818.40 | 222.00 | 3287.20 | 16003.00 | 868.00 | 15965.40 | 802.00 | 15947.20 |
| 5 | 712 | 120 | 7140 | 0.30 | 0.30 | 5819.40 | 187.40 | 5536.60 | 15372.00 | 871.80 | 15306.40 | 712.00 | 15790.40 |
| | | | | 0.50 | 0.50 | 5717.20 | 293.80 | 5141.00 | 31324.80 | 805.40 | 30781.40 | 712.00 | 31503.20 |
| | | | | 0.70 | 0.70 | 5801.40 | 316.00 | 5035.20 | 47519.00 | 804.20 | 47047.20 | 712.00 | 47865.80 |
| 6 | 793 | 160 | 12720 | 0.30 | 0.30 | 7434.80 | 284.40 | 7050.40 | 41993.60 | 1353.60 | 42418.60 | 809.60 | 42628.20 |
| | | | | 0.50 | 0.50 | 7361.00 | 421.80 | 7111.60 | 87118.80 | 1061.60 | 86987.40 | 793.00 | 86828.40 |
| | | | | 0.70 | 0.70 | 7517.00 | 403.20 | 6735.00 | 163025.00 | 955.40 | 161862.40 | 793.00 | 154731.20 |

avg.: average solution of 20 runs; CPU time: average computation time in millisecond (ms).

The experimental study was realized to investigate the effectiveness of the different encoding method; the interaction of the encoding with the crossover operators and mutation operators, and the parameter settings affect its performance. Table 8 gives computational results for four different encoding methods on six test problems by three kinds of parameter settings. In the columns of the best cost of four encoding methods, it is possible to see that whereas the Prüfer number-based approach is faster than the others, it is difficult to build from the substructures of their parents' phenotypes (poor heritability), and the result is very far from the best one. Two kinds of mutation are used in edge-based encoding, the second one (depends on the cost) giving better performance than the first. For considering the computational cost (CPU time), because of the LowestCost mutation in the proposed approach, spending a greater CPU time to find the edge with the lowest cost they always longer than other algorithms. However, PrimPred-based GA developed in this study gives a better cost than other algorithms.

Then we show performance comparisons of multiobjective GAs for solving bSTP by different fitness assignment approaches, there are spEA, nsGAII, rwGA and i-awGA. The data in test problem was generated randomly. In each GA approach, PrimPred-based encoding was used, and Prim-based crossover and LowestCost mutation were used as genetic operators. GA parameter settings were taken as follows:

Population size: *popSize* =20;
Crossover probability: $p_C = 0.70$;
Mutation probability: $p_M = 0.50$;
Stopping criteria: evaluation of 5000 solutions

We compare i-awGA with spEA, nsGAII and rwGA trough computational experiments on the 40-node/1560-arc test problem under the same stopping condition (*i.e.*, evaluation of 5000 solutions). Each algorithm was applied to each test problem 10 times and gives the average results of the 3 performance measures (*i.e.*, the number of obtained solutions $|S_j|$, the ratio of nondominated solutions $R_{NDS}(S_j)$, and the average distance $D1_R$ measure). In Table 9, better results of all performance measures were obtained from the i-awGA than other fitness assignment approach.

**Table 9.** Performance Evaluation of Fitness Assignment Approaches for the 40-node/1560-arc Test Problem

| # of | $\|S_j\|$ | | | | $R_{NDS}(S_j)$ | | | | $D1_R(S_j)$ | | | |
|---|---|---|---|---|---|---|---|---|---|---|---|---|
| eval. solut. | spEA | nsGAII | rwGA | i-awGA | spEA | nsGAII | rwGA | i-awGA | spEA | nsGAII | rwGA | i-awGA |
| 50 | 31.45 | 30.40 | 32.60 | 36.20 | 0.34 | 0.31 | 0.36 | 0.39 | 178.85 | 200.47 | 182.03 | 162.57 |
| 500 | 42.40 | 45.60 | 43.20 | 47.60 | 0.42 | 0.45 | 0.40 | 0.52 | 162.97 | 151.62 | 160.88 | 157.93 |
| 2000 | 46.60 | 52.20 | 45.30 | 55.50 | 0.54 | 0.61 | 0.58 | 0.66 | 118.49 | 114.60 | 139.40 | 92.41 |
| 5000 | 51.20 | 54.40 | 50.30 | 60.70 | 0.64 | 0.70 | 0.62 | 0.73 | 82.70 | 87.65 | 117.48 | 77.98 |

## 4.4 Experiments for bNF Model

In the experimental study, we demonstrate the performance comparisons of multiobjective GAs for solving bNF problems by different fitness assignment approaches. We compare i-awGA with spEA, nsGAII and rwGA trough computational experiments on the 25-node/49-arc and 25-noed/56-arc test problems [29] under the same GA parameter settings: population size, *popSize* =20; crossover probability, $p_C$=0.70; mutation probability, $p_M$ =0.70; stopping condition, evaluation of 5000 solutions.

The number of the obtained reference solutions for 2 test problems is summarized in Table 10. We chose nondominated solutions as reference solutions from 4 solution sets of the four algorithms for each test problem. We show the obtained reference solution sets for the 25-node / 49-arc test problem in Fig. 1(a), 25-noed / 56-arc test problem in Fig. 1(b), respectively. We can observe the existence of a clear tradeoff between the two objectives in each figure. We can also see that the obtained reference solution set for each test problem has a good distribution on the tradeoff front in the objective space.

**Table 10.** Number of obtained reference solutions and their range width for each objective

| Test Problems (# of nodes / # of arcs) | # of obtained solutions $\|S_j\|$ | range width $W_{f_i}(S^*)$ | |
|---|---|---|---|
| | | $f_1(r)$ | $f_2(r)$ |
| 25 / 49 | 69 | 85 | 19337 |
| 25 / 56 | 77 | 89 | 16048 |

where, the range width of the $i$th objective over the reference solution set $S^*$ is defined as:

$$W_{f_i}(S^*) = \max\{f_i(\boldsymbol{r})|\boldsymbol{r} \in S^*\} - \min\{f_i(\boldsymbol{r})|\boldsymbol{r} \in S^*\}$$

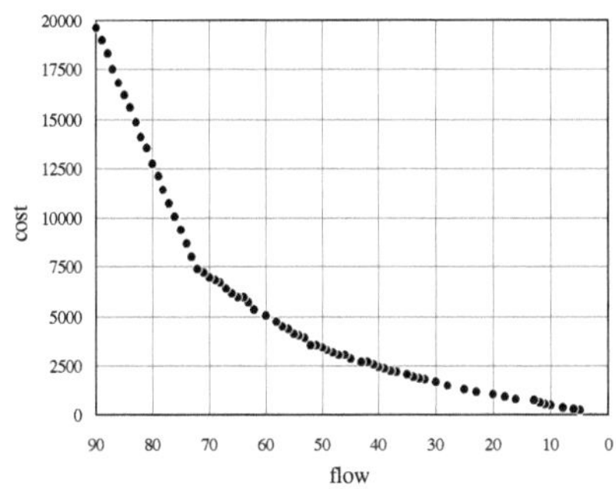

(a) 25-node/49-arc test problem

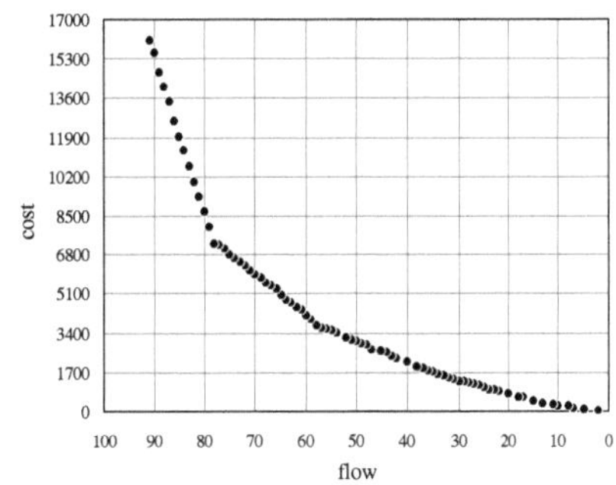

(b) 25-node/56 arc test problem

**Fig. 1.** Reference solutions obtained from the four GA approaches

Each algorithm was applied to each test problem 10 times and gives the average results of the 3 performance measures (*i.e.*, the number of obtained solutions $|S_j|$, the ratio of nondominated solutions $R_{NDS}(S_j)$, and the average distance $D1_R$ measure). In Table 11, better results of $|S_j|$ and $D1_R$ were obtained from the i-awGA than other

**Table 11.** Performance Evaluation of Fitness Assignment Approaches for the 25-node/49-arc Test Problem

| # of eval. solut. | $\|S_j\|$ spEA | nsGAII | rwGA | i-awGA | $R_{NDS}(S_j)$ spEA | nsGAII | rwGA | i-awGA | $D1_R(S_j)$ spEA | nsGAII | rwGA | i-awGA |
|---|---|---|---|---|---|---|---|---|---|---|---|---|
| 50 | 41.60 | 40.60 | 40.30 | 42.40 | 0.44 | 0.42 | 0.45 | 0.49 | 201.25 | 210.63 | 205.03 | 184.12 |
| 500 | 51.40 | 56.30 | 49.40 | 54.60 | 0.54 | 0.60 | 0.53 | 0.64 | 151.82 | 124.81 | 149.44 | 132.93 |
| 2000 | 58.20 | 60.60 | 54.30 | 59.20 | 0.62 | 0.71 | 0.62 | 0.75 | 108.49 | 101.45 | 127.39 | 88.99 |
| 5000 | 60.70 | 61.60 | 58.30 | 61.40 | 0.72 | 0.80 | 0.67 | 0.82 | 79.91 | 80.70 | 103.70 | 67.14 |

**Table 12.** Performance Evaluation of Fitness Assignment Approaches for the 25-node/56-arc Test Problem

| # of | $\lvert S_j \rvert$ | | | | $R_{NDS}(S_j)$ | | | | $D1_R(S_j)$ | | | |
|---|---|---|---|---|---|---|---|---|---|---|---|---|
| eval. solut. | spEA | nsGAII | rwGA | i-awGA | spEA | nsGAII | rwGA | i-awGA | spEA | nsGAII | rwGA | i-awGA |
| 50 | 41.20 | 43.60 | 42.60 | 44.00 | 0.35 | 0.33 | 0.34 | 0.33 | 181.69 | 180.64 | 168.73 | 168.96 |
| 500 | 49.80 | 56.60 | 51.60 | 57.50 | 0.47 | 0.50 | 0.42 | 0.46 | 104.77 | 114.62 | 119.53 | 103.13 |
| 2000 | 62.90 | 62.90 | 55.30 | 64.70 | 0.61 | 0.65 | 0.51 | 0.65 | 74.76 | 81.24 | 95.70 | 76.41 |
| 5000 | 67.80 | 68.40 | 60.70 | 69.40 | 0.73 | 0.72 | 0.64 | 0.73 | 62.97 | 62.77 | 80.68 | 62.33 |

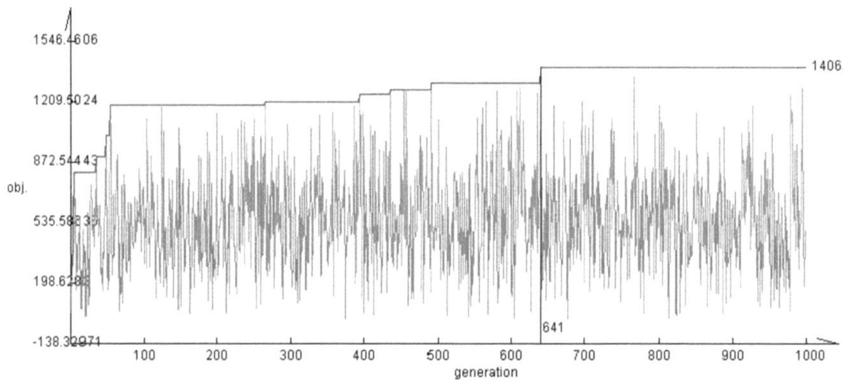


**Fig. 2.** Evolutionary process on 25-node/49-arc test problem

fitness assignment approach. The results of $R_{NDS}(S_j)$ are no large differences among the 4 fitness assignment approaches. In Table 12, better results of all performance measures were obtained from the i-awGA than other fitness assignment approach. An evolutionary process of the fitness by using our i-awGA for solving the 25-node/49-arc test problem is shown in Fig. 2.

# 5 Conclusions

In this paper, we investigated with a broad spectrum of bicriteria network optimization models, analyze the recent related researches, design and validate effective EA approaches for the typical network models: bicriteria shortest path (bSP) model, bicriteria minimum spanning tree (bMST) model, bicriteria network flow (bNF) model. Focus on the broad applications of proposed evolutionary algorithms (EAs) to network design with large-scale optimization.

For bSP model, we introduced an efficient genetic representation using the priority-based encoding method. We gave effective analysis of different evolutionary approaches for bSP model dedicated to calculate nondominated paths for the minimum total cost and the minimum transmission delay. For bST model, we investigated different GA approaches for solving minimum spanning tree (MST) problems, and introduced a new genetic representation using the PrimPred-based encoding method.

For bNF model, we introduced a new multiobjective genetic algorithm (moGA) to solve the problem with two conflicting objectives to minimize the total cost and maximize the total flow simultaneously.

**Acknowledgments.** This work is partly supported by the Ministry of Education, Science and Culture, the Japanese Government: Grant-in-Aid for Scientific Research (No.19700071, No. 20500143).

## References

1. Ahuj, R.K., Magnanti, T.L., Orlin, J.B.: Network Flows, New Jersey (1993)
2. Garey, M.R., Johnson, D.S.: Computers and Intractability: a guide to the theory of NPcompleteness. W. H. Freeman, San Francisco (1979)
3. Hansen, P.: Bicriterion path problems. In: Proceeding 3rd Conference Multiple Criteria Decision Making Theory and Application, pp. 109–127 (1979)
4. Skriver, A.J.V., Andersen, K.A.: A label correcting approach for solving bicriterion shortest-path problems. Computers & Operations Research 27(6), 507–524 (2000)
5. Azaron, A.: Bicriteria shortest path in networks of queues. Applied Mathematics & Comput. 182(1), 434–442 (2006)
6. Marathe, M.V., Ravi, R., Sundaram, R., Ravi, S.S., Rosenkrantz, D.J., Hunt, H.B.: Bicriteria network design problems. Journal of Algorithms 28(1), 142–171 (1998)
7. Balint, V.: The non-approximability of bicriteria network design problems. Journal of Discrete Algorithms 1(3,4), 339–355 (2003)
8. Lee, H., Pulat, P.S.: Bicriteria network flow problems: continuous case. European Journal of Operational Research 51(1), 119–126 (1991)
9. Yuan, D.: A bicriteria optimization approach for robust OSPF routing. Proceeding IEEE IP Operations & Management, 91–98 (2003)
10. Yang, H., Maier, M., Reisslein, M., Carlyle, W.M.: A genetic algorithm-based methodology for optimizing multiservice convergence in a metro WDM network. J. Lightwave Technol. 21(5), 1114–1133 (2003)
11. Raghavan, S., Ball, M.O., Trichur, V.: Bicriteria product design optimization: an efficient solution procedure using AND/OR trees. Naval Research Logistics 49, 574–599 (2002)
12. Zhou, G., Min, H., Gen, M.: A genetic algorithm approach to the bi-criteria allocation of customers to warehouses. International Journal of Production Economics 86, 35–45 (2003)
13. Gen, M., Cheng, R., Oren, S.S.: Network Design Techniques using Adapted Genetic Algorithms. Advances in Engineering Software 32(9), 731–744 (2001)
14. Ahn, C.W., Ramakrishna, R.: A genetic algorithm for shortest path routing problem and the sizing of populations. IEEE Transaction on Evolutionary Computation 6(6), 566–579 (2002)
15. Wu, W., Ruan, Q.: A gene-constrained genetic algorithm for solving shortest path problem. In: Proceeding 7th International Conference Signal Processing, vol. 3, pp. 2510–2513 (2004)
16. Li, Q., Liu, G., Zhang, W., Zhao, C., Yin, Y., Wang, Z.: A specific genetic algorithm for optimum path planning in intelligent transportation system. In: Proceeding 6th International Conference ITS Telecom, pp. 140–143 (2006)
17. Kim, S.W., Youn, H.Y., Choi, S.J., Sung, N.B.: GAPS: The genetic algorithm based path selection scheme for MPLS network. In: Proceeding of IEEE International Conference on Information Reuse & Integration, pp. 570–575 (2007)

18. Hasan, B.S., Khamees, M.A., Mahmoud, A.S.H.: A heuristic genetic algorithm for the single source shortest path problem. In: Proceeding IEEE/ACS International Conference on Computer Systems & Applications, pp. 187–194 (2007)
19. Ji, Z., Chen, A., Subprasom, K.: Finding multi-objective paths in stochastic networks: a simulation-based genetic algorithm approach. In: Proceedings of IEEE Congress on Evolutionary Computation, vol. 1, pp. 174–180 (2004)
20. Chakraborty, B., Maeda, T., Chakraborty, G.: Multiobjective route selection for car navigation system using genetic algorithm. In: Proceeding of IEEE Systems, Man & Cybernetics Society, pp. 190–195 (2005)
21. Garrozi, C., Araujo, A.F.R.: Multiobjective genetic algorithm for multicast routing. In: Proceeding IEEE Congress on Evolutionary Computation, pp. 2513–2520 (2006)
22. Kleeman, M.P., Lamont, G.B., Hopkinson, K.M., Graham, S.R.: Solving multicommodity capacitated network design problems using a multiobjective evolutionary algorithm. In: Proceeding IEEE Computational Intelligence in Security & Defense Applications, pp. 33–41 (2007)
23. Gen, M., Cheng, R.: Genetic Algorithms and Engineering Optimization. John Wiley & Sons, New York (2000)
24. Bazaraa, M., Jarvis, J., Sherali, H.: Linear Programming and Network Flows, 2nd edn. John Wiley & Sons, New York (1990)
25. Munemoto, M., Takai, Y., Sate, Y.: An adaptive network routing algorithm employing path genetic operators. In: Proceeding of the 7th International Conference on Genetic Algorithms, pp. 643–649 (1997)
26. Inagaki, J., Haseyama, M., Kitajim, H.: A genetic algorithm for determining multiple routes and its applications. In: Proceeding of IEEE International Symposium. Circuits and Systems, pp. 137–140 (1999)
27. Cheng, R., Gen, M.: Evolution program for resource constrained project scheduling problem. In: Proceedings of IEEE International Conference of Evolutionary Computation, pp. 736–741 (1994)
28. Gen, M., Cheng, R., Wang, D.: Genetic algorithms for solving shortest path problems. In: Proceedings of IEEE International Conference of Evolutionary Computation, pp. 401–406 (1997)
29. Lin, L., Gen, M.: Bicriteria network design problem using interactive adaptive-weight GA and priority-based encoding method. IEEE Transactions on Evolutionary Computation in Reviewing (2007)
30. Davis, L., Orvosh, D., Cox, A., Qiu, Y.: A genetic algorithm for survivable network design. In: Proceeding 5th International Conference Genetic Algorithms, pp. 408–415 (1993)
31. Piggott, P.I., Suraweera, F.: Encoding graphs for genetic algorithms: an investigation using the minimum spanning tree problem. In: Yao, X. (ed.) AI-WS 1993 and 1994. LNCS (LNAI), vol. 956, pp. 305–314. Springer, Heidelberg (1995)
32. Bean, J.C.: Genetic algorithms and random keys for sequencing and optimization. ORSA J. Computing 6(2), 154–160 (1994)
33. Schindler, B., Rothlauf, F., Pesch, H.: Evolution strategies, network random keys, and the one-max tree problem. In: Proceeding Application of Evolutionary Computing on EvoWorkshops, pp. 143–152 (2002)
34. Rothlauf, F., Gerstacker, J., Heinzl, A.: On the optimal communication spanning tree problem. IlliGAL Technical Report, University of Illinois (2003)
35. Knowles, J., Corne, D.: A new evolutionary approach to the degree-constrained minimum spanning tree problem. IEEE Transaction Evolutionary Computation 4(2), 125–134 (2000)

36. Raidl, G.R., Julstrom, B.: Greedy heuristics and an evolutionary algorithm for the bounded-diameter minimum spanning tree problem. In: Proceeding SAC, pp. 747–752 (2003)
37. Cayley, A.: A theorem on tree. Quarterly Journal of Mathematics & Physical Sciences 23, 376–378 (1889)
38. Prüfer, H.: Neuer bewis eines Satzes über Permutationnen. Archives of Mathematical Physica 27, 742–744 (1918)
39. Schaffer, J.D.: Multiple objective optimization with vector evaluated genetic algorithms. In: Proc. 1st Inter. Conf. on GAs, pp. 93–100 (1985)
40. Fonseca, C., Fleming, P.: Genetic algorithms for multiobjective optimization: formulation, discussion and generalization. In: Proc. 5th Inter. Conf. on Genetic Algorithms, pp. 416–423 (1993)
41. Srinivas, N., Deb, K.: Multiobjective function optimization using nondominated sorting genetic algorithms. Evolutionary Computation 3, 221–248 (1995)
42. Ishibuchi, H., Murata, T.: A multiobjective genetic local search algorithm and its application to flowshop scheduling. IEEE Trans. on Systems., Man & Cyber. 28(3), 392–403 (1998)
43. Zitzler, E., Thiele, L.: SPEA2: improving the strength Pareto evolutionary algorithm, Technical Report 103, Computer Engineering and Communication Networks Lab, TIK (2001)
44. Raidl, G.R., Julstrom, B.A.: Edge Sets: An Effective Evolutionary Coding of Spanning Trees. IEEE Transaction on Evolutionary Computation 7(3), 225–239 (2003)
45. OR-Library, http://people.brunel.ac.uk/mastjjb/jeb/info.html
46. Lin, L., Gen, M.: An effective evolutionary approach for bicriteria shortest path routing problems. IEEJ Transactions on Electronics, Information and Systems 128(3), 416–443 (2008)
47. Deb, K., Pratap, A., Agarwal, S., Meyarivan, T.: A fast and elitist multiobjective genetic algorithm: NSGA-II. IEEE Trans. Evolutionary Computation 6(2), 182–197 (2002)

# Use of Serendipity Power for Discoveries and Inventions

Shigekazu Sawaizumi, Osamu Katai, Hiroshi Kawakami, and Takayuki Shiose

Graduate School of Informatics, Kyoto University,
Yoshida-Honmachi, Sakyo-ku, Kyoto 606-8501, Japan
sawaizumi@sys.i.kyoto-u.ac.jp,
{katai,kawakami,shiose}@i.kyoto-u.ac.jp

**Abstract.** The word "serendipity," introduced to scientific fields by R. K. Merton makes discoveries by accidents and sagacity and has long life since the eighteenth century. Its power was experimentally studied for education in scientific observations by R. S. Lenox. In this paper, in a simple model we analyzed the power of serendipity with two important factors: accident and sagacity. A method to improve its power is also presented based on a mechanism of a serendipity phenomenon that works effectively.

**Keywords:** Serendipity, discovery, accident, sagacity, serendipity card, brain inventory, small world, strength of ties.

## 1 Introduction

R. K. Merton wrote that serendipity resonated more for him because Walpole defined this "very expressive" word as referring to "discoveries, by accidents and sagacity [1]." The word "serendipity" coined by H. Walpole contains two keywords "accident" and "sagacity" [2] for making discoveries.

Serendipity not only follows the name of the Kingdom of Serendip but also implies "Seren" that means "Serene" or "His Serene Highness," "-dip-" that means "dip (something important)" and "-ty" of "-(i)ty" that means both quality and power. In this interpretation, we added new reading of "Seren" in the references. Even though the power of serendipity has been studied by many persons, it is not completely used yet or it is still in vain.

In this study, we analyze the roles of accidents and sagacity, even their synergy effect. By studying the power of serendipity, we have developed a method and a tool incorporated in a serendipity card system to use its power and to plan future assessment. The serendipity card system is based on our theory in a simple model on the work of serendipity.

We are planning to introduce serendipity power to increase the chances of discoveries in many divisions of organizations and to develop an assessment system using serendipity cards.

## 2 Serendipity and Its Effect

Discoveries often happen in unexpected circumstances. The role of this phenomenon of discovery is related to the current paradigm. As Kuhn wrote, "Normal science does

M. Gen et al.: Intelligent and Evolutionary Systems, SCI 187, pp. 163–169.
springerlink.com 

not aim at novelties of fact or theory and when successful, find none [3]." A discovery in a planned program is also difficult to make an advanced innovation with the restriction of the current paradigm. Because the idea of planned discovery in the current paradigm usually falls within the expected results, but the idea of accidental discovery is not restricted by the current paradigm. This means that a serendipitous discovery sometimes causes a very important result that gives birth to a new paradigm. Accidents can be systematically increased by planned actions, and the faculty of sagacity can also be improved by training. We developed a system to increase encounters by making hypotheses and to refresh memories by the "brain inventory." The method proposed by R. Lenox [4] is also appropriate for making serendipitous discoveries in observations.

We propose that the power of serendipity is useful in many fields such as academic research, a planning of projects and a development of business models. For applying the power of serendipity efficiently, we use a serendipity card system.

In this paper, we define "serendipity" as "the faculty of making happy and unexpected discoveries by accident" and define "serendipity phenomenon" in the meaning of "the act of something interesting or pleasant happening by chance."

## 3 Role of Accidents and Their Power

Accidents in serendipity loosen the force of the current paradigm. As Kuhn wrote, "The discoveries come out when the old paradigm has to change." The idea of discovery is usually restrained by the current paradigm. An accident is free from this current paradigm since it allows no time to be considered by them [5].

Even the ratio of the success of the serendipity phenomenon is much low, an attractive point of the phenomenon may completely change the current paradigm. Another role of an accident may cause uneven distribution. During times of better distribution, we are ready to "catch" a better condition to obtain solution. We may attain a solution by amplifying uncertainty [6].

The numbers of encounters may be increased by making hypotheses, because we are more concerned with a subject whose result we predicted. In this era of information technology, we may use more effectively information encounters through computers. D. Green wrote that "Combining lots of different pieces of data leads inevitably to serendipity, to unexpected discoveries. Some of these discoveries are trivial. But even only one in a thousand combinations leads to some unexpected discovery; the potential number of discoveries – unexpected, serendipitous discoveries – is still enormous" [7].

## 4 Role of Sagacity and Its Power

Occasions of effective use of serendipity are shown in Fig. 1 symbolically by a simple model of a knowledge level in a certain field. In Fig. 1, the vertical axis shows the intensity level and the horizontal axis shows the time, and the lines are defined as follows:

Point-A: Time when a phenomenon was observed.
Point-B: Time when the phenomenon was acknowledged.

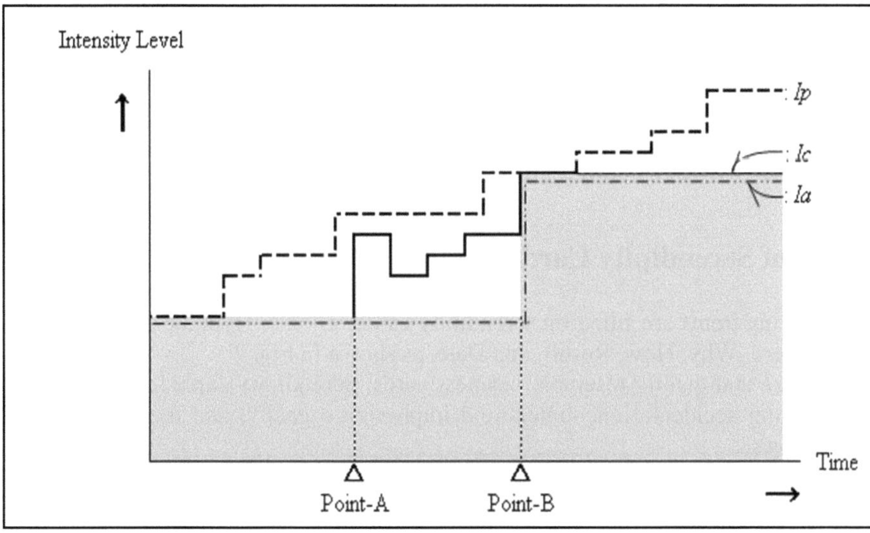


**Fig. 1.** A simple model of a serendipitous discovery is shown symbolically with intensity levels of a phenomenon and an inferred level. In the case that a phenomenon is noticed exceeding an acknowledged level in the current paradigm, hypotheses are developed into discovery between Point-A and Point-B.

*Ip*: Intensity of a phenomenon
*Ic*: Inferred level of the phenomenon
*Ia*: Acknowledged level in the current paradigm

One notices *Ip* for a discovery about an encounter of phenomenon, in the case that one understands it exceeding *Ia*. A notice to have acuteness of mental discernment like sagacity is required to break through *Ia* for making a discovery.

Sagacity creates serendipity in the following case shown in an equation as a discovery of serendipity pattern:

$$| Ip - Ia | \gg | Ip - Ic | \quad (1)$$

For example, for the discovery of X rays, the significance noticing of the difference between "the phenomenon in which a ray penetrate a body" and "the knowledge in which a ray does not penetrate a body" makes a chance for the discovery.

The reason for the necessity *Ip* is that the difference between *Ip* and *Ia* is prone to be disregarded by a person who understands a phenomenon within his experience.

On the other hand, one who knows the significance of this break through mechanism can lower the right hand side of the equation, as a specialist of a certain field can notice in a different field with curiosity. In this case, we adjust the equation with a personal factor *k* to notice for discover.

$$| Ip - Ia | \gg k\, | Ip - Ic | \quad (2)$$

When *Ip* is taken for *Ic* by mistake, sagacity works for serendipity. This occasion is identity, as shown in "The force of Falsity" by U. Eco [9]. Eco said that "in the field of science, this mechanism is known as serendipity" and provided historical example of the voyages of Columbus. This model also shows that the richness of experience does not always work well to create an innovative idea under the restrains of the current paradigm.

Since *Ic* has a tendency to follow the value of *Ia* in a current paradigm, it is effective to externalize *Ic* on serendipity cards for practicing serendipitous discoveries.

## 5 Role of Serendipity Card

The following items are filled on serendipity cards: Theme, Hypothesis, Who, What, When, Where, Why, How, Result, and Date, as shown in Fig. 2.

Their size and quality resemble business cards. Serendipity cards have many roles for increasing accidental encounters and improving sagacity, and their main effects are as follows:

(1) For increasing encounters with accidents:
- To make a theme for clarifying a subject
- To make an hypothesis for obtaining related information
- To make related items 5W + 1H for clusters and hubs
- To make a model of creativity verification
- To refresh memories by the brain inventory

(2) For improving sagacity:
- To externalize an idea from the brain for visualization by card
- To make clusters based on themes and items for making encounters of ideas
- To make hubs based on items for making a short pass of the "small world"
- To merely dip into an interesting items
- To make different clusters with identical cards to change perspective

We fill out a card each time we get an idea of hypothesis and we use these cards when we need a new idea for some projects. With about thirty cards, we make clusters with similar themes and then remake different clusters with similar items using the identical cards. In the movement of cards in these clusters, we may stumble on a new idea and have a chance to make new cards. We named the flow of these related action the "brain inventory" [8].

D. Chambers recognized that filling in cards means to externalize ideas from the brain, and studied the effect of this action [9]. The serendipity card system refers to the KJ-method by J. Kawakita, the NM-method by M. Nakayama, and the Kyoto University-card system by T. Umesada since they have studied observation and thinking with these systems [10], [11], [12].

By studying the brain inventory, we recognized that the concept of "small world" is useful to make effective clusters, nodes, and paths for the brain inventory or a serendipity card system. For developing an effective brain inventory, clusters with short paths are important for the association of ideas.

The effect of encounters of ideas is related to the theory of the "strength of weak ties" [13], because unexpected interpersonal ties are related to the encounters of ideas.

It is applied to increase numbers of effective encounters for serendipitous discoveries, since the theory of the "strength of weak ties" associate with other specified fields,

## 6 Example of Using Serendipity Card System

As shown in Fig. 2, the serendipity card size limits its amount of information. However, the externalization of ideas from the brain is crucial, and the process of considering a hypothesis is significant for reminding subjects of encounters with related stimuli. The real nature of the serendipity card system is to make unexpected encounters among many hypotheses over a long period. A few serendipity cards may not result in an encounter of discovery, but as cards accumulate, they make encounters with a synergy effect of power.

An example of the use of serendipity cards is introduced on invitations to an international workshop in Toyama prefecture. Even though it was difficult to foresee the result due to the expectancy or the advertisement of other fields, a connection of related person made it possible. Such a solution is quite often observed when we are keen to prepare cards in a daily occurrence.

In the beginning of the plan, invitations did not meet with organizations approval due to disadvantage when compared with other big cities. Complicated associations of serendipity cards suggested that a few organizations may support a host to hold the workshop, due to promotion campaign in the prefecture.

In the case concerned with this example, in ten months we encountered more than thirty unexpected information that supported to hold the workshop in Toyama prefecture.

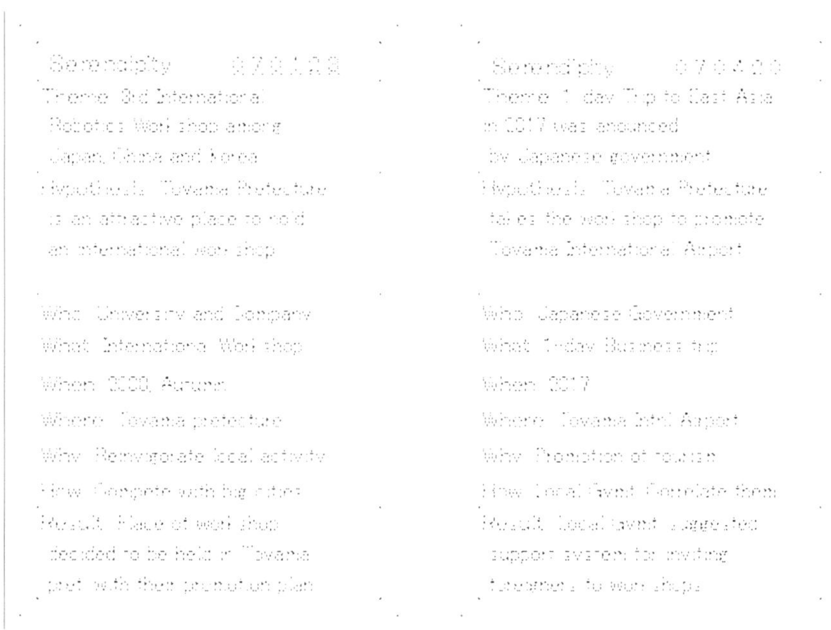


**Fig. 2.** An Example of Encountered Two Serendipity Cards

The result shows that a solution was found from other points of view for other purpose. To obtain a chance to use an unexpected encounter, we must be ready to remind ourselves of our goals. For expanding a chance, it is recommended to show the purpose and difficulties with a subject for receiving an advice from specialized persons. Additionally, since some information has a searching direction, it is more difficult to find it from the receiving side than the transmitting side.

The recognition that unexpected encounters with different specialized fields bring a solution has great significance for the serendipity card system. We evaluate the effect of accumulated serendipity cards to find unexpected encounters. One often gives up too easily for breaking a precedent of special fields, but changes or advancement in the world make their breaking through possible in every day.

Our proposed system is to hypothesize subjects on serendipity cards, to collect related information, to associate cards in the brain inventory system, to verify a subject by abduction method, and to discover a subject with significant roles of serendipity cards.

For researching their effectiveness, we study on several clusters of externalized idea in hypotheses. We remind ourselves of the ideas of subjects by arranging clusters, as each cluster consists of an identical theme in some case and an identical 5W+1H item in other case.

A flowchart of causing a typical serendipitous discovery with cards is shown in Fig. 3.

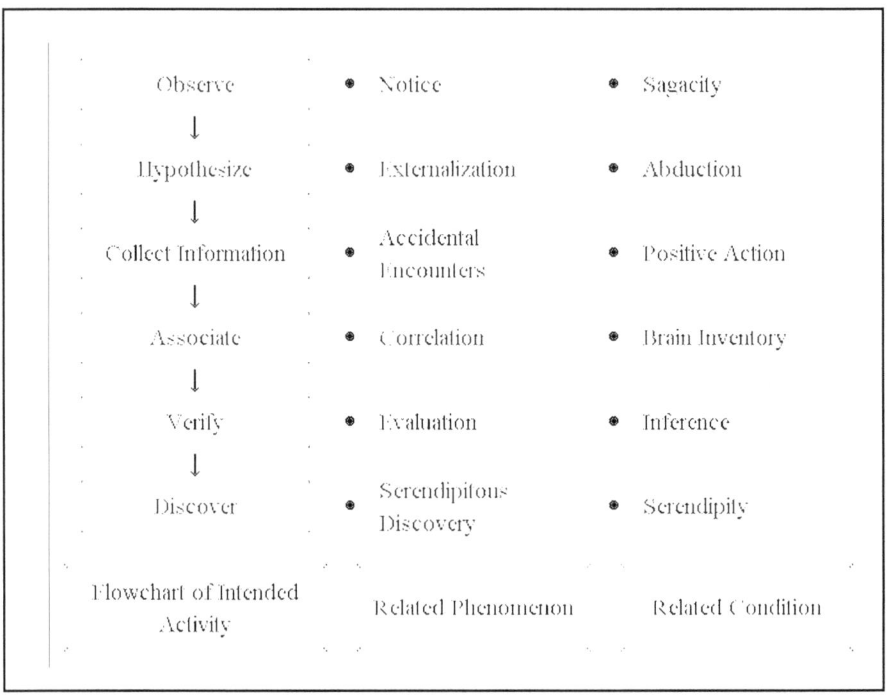


**Fig. 3.** Flowchart of Serendipitous Discovery

## 7 Further Study

By studying a simple model of serendipitous phenomenon, we recognized several types of serendipitous discoveries. Some of them appear within the target of the discoverer and others appear unexpectedly. The former is called pseudo serendipity due to their expected appearance, and it has been usually greatly pre-studied about the subject. We are interested in the difference between serendipity and pseudo serendipity, because their reasons will be useful for further study of hypotheses.

We understand the necessity of making an assessment to count the effect of using serendipity cards. First, the numbers of accidental encounters in the study may be counted as we count the numbers of records of hypothesis in cards.

We also consider how to think through serendipity, especially since we have too many factors or causes in each discipline to find a phenomenon in today's science.

However, in some cases, we need a more macroscopic view over different disciplines.

This is the way of serendipity: to see a phenomenon in very simple factors, such as by accident and sagacity. We would like to develop useful serendipity card system that is supported with intelligent theory.

## References


1. Merton, R.K., Barber, E.: The Travels and Adventures of Serendipity. Princeton University Press, Princeton (2004)
2. Lewis, W.S. (ed.): The Yale Edition of Horace Walpole's Correspondence, vol. 20, pp. 407–411 (1960)
3. Kuhn, T.S.: The Structure of Scientific Revolutions, 3rd edn., p. 52. The University of Chicago Press (1996)
4. Lenox, R.S.: Educating for the Serendipitous Discovery. Journal of Chemical Education 62(4), 282–285 (1985)
5. Murakami, Y., Ohtani, T.: The Social Basis of Scientific Discoveries, by Augustine Branningan, p. J63. Cambridge University Press, Kinokuniya Shoten (1984)
6. Hioki, K.: Nihongata MOT (MOT in Japanese Style), Chuo Keizaisha (2005) (in Japanese)
7. Green, D.: The Serendipity Machine. Allen & Unwin (2004)
8. Sawaizumi, S., Shiose, T., Katai, O., Kawakami, H.: A consideration on modeling of sparks under the specific condition -The use of serendipity effect. In: Proc. of 32th SICE (2005)
9. Eco, U.: Serendipities: Language and Lunacy. Translated by William Weaver. Orion Paperback (2005)
10. Chambers, D., Reisber, D.: What an image depicts depends on what an image means. Cognitive Psychology 24, 145–174 (1995)
11. Kawakita, J.: Hassoho (The way of thinking). Chuukou Shinsho (in Japanese) (1967)
12. Nakayama, M.: Hasso no Ronri (The Logic of Making Idea). Chuukou Shinsho (in Japanese) (1970)
13. Umesao, T.: Chiteki Seisan no Gijutsu (The Technique of Intellectual Production). Iwanami Shinsho (in Japanese) (1969)
14. Granovetter, M.S.: The Strength of Weak Ties. American Journal of Sociology 78(6), 1360–1379 (1972)

# Evolution of Retinal Blood Vessel Segmentation Methodology Using Wavelet Transforms for Assessment of Diabetic Retinopathy

D.J. Cornforth[1], H.F. Jelinek[2], M.J. Cree[3], J.J.G. Leandro[4], J.V.B. Soares[4], and R.M. Cesar Jr.[4]

[1] School of Information Technology and Electrical Engineering, University of New South Wales, ADFA, Canberra, Australia
d.cornforth@adfa.edu.au
[2] School of Community Health, Charles Sturt University, Albury, NSW, Australia
hjelinek@csu.edu.au
[3] Dept. Engineering, University of Waikato, Hamilton, New Zealand
cree@waikato.ac.nz
[4] Computer Science, University of São Paulo, Brazil
{jleandro,joao,cesar}@vision.ime.usp.br

## 1 Introduction

Diabetes is a chronic disease that affects the body's capacity to regulate the amount of sugar in the blood. One in twenty Australians are affected by diabetes, but this figure is conservative, due to the presence of subclinical diabetes, where the disease is undiagnosed, yet is already damaging the body without manifesting substantial symptoms. This incidence rate is not confined to Australia, but is typical of developed nations, and even higher in developing nations. Excess sugar in the blood results in metabolites that cause vision loss, heart failure and stroke, and damage to peripheral blood vessels. These problems contribute significantly to the morbidity and mortality of the Australian population, so that any improvement in early diagnosis would therefore represent a significant gain. The incidence is projected to rise, and has already become a major epidemic [16].

The most common diagnostic test for diabetes is measurement of blood sugar, but this is only effective when the disease has already made substantial progression. However, because of the effect of diabetes on peripheral vessels, it is possible to detect diabetes by examining these vessels. One of the most suitable areas to make such an observation is the retina, where small blood vessels are arranged on the surface, and visual inspection is possible through the pupil itself. This technique is well developed, with ophthalmologists routinely employing manual inspection of the retina for diagnosing diabetic retinopathy, which is caused by diabetes, and leads to significant vision degeneration without prompt treatment. In addition cameras can capture an image of the retina for examination by ophthalmologists or for telemedicine as well as for providing records over time. The requirement of specialists to make an accurate diagnosis does make retinal

M. Gen et al.: Intelligent and Evolutionary Systems, SCI 187, pp. 171–182.
springerlink.com 

photography prohibitive in cost as a screening tool for the general population, especially in rural or remote regions.

Images containing labelled blood vessels can be derived by injecting a fluorescent dye into the person being examined, so that blood vessels can be observed with higher contrast. This technique, know as fluorescein imaging, is invasive and brings some risk. As it also requires the presence of an ophthalmologist, it is not suitable for rural and remote screening programmes. Images taken without fluorescent dye and pupil dilation are known as non-mydriatic, and are also less invasive with good contrast due to the high resolution cameras available. These are therefore desirable for use in remote or rural areas as they can be obtained by trained rural health professionals such as indigenous health workers, diabetes educators and community nurses.

The aim of this work is first, to improve the accuracy and speed of vessel segmentation using non-mydriatic retinal images, by the application of advanced image processing techniques; and second, to apply machine intelligence techniques to offer decision support and reduce the burden on specialist interpretation. Starting with a non-mydriatic image, our aim is to provide an assessment of risk of diabetes for the person being examined.

Identification of anomalies in retinal blood vessels, associated with diabetes health care, represents a large portion of the assessment carried out by ophthalmologists, which is time consuming and in many cases does not show any anomalies at the initial visit. Utilizing non-specialist health workers in identifying diabetic eye disease is an alternative but trials have shown that correct identification of retinal pathology may be poor (i.e. only 50% of the cases). This success rate decreases for early proliferative retinopathy stages. Telemedicine is an attractive approach. However, this procedure is not time effective and does not lessen the burden on a comparatively small number of ophthalmologists in rural areas that need to assess the images. In addition significant technical problems lessen the availability of telemedicine [21].

## 2 Image Processing for Medical Diagnosis

Automated assessment of blood vessel patterns that can be used by rural health professionals is now being extended from fluorescein-labelled to non-mydriatic camera images [3, 15]. This has the advantage of a less invasive and risky procedure, making possible a screening procedure for the general population. A significant problem in these non-mydriatic images, however, is the ability to identify the blood vessels in low vessel to background contrast and diverse pathology, and to separate (segment) them from the background image (fundus). In this work we present the evolution of retinal blood vessel segmentation, using the wavelet transform combined with mathematical morphology, supervised training algorithms and adaptive thresholding. Once the vessels have been successfully segmented, it is possible to apply automated measures, such as morphology measures, then to use further automated methods to identify anomalies. This further processing is outside the scope of this paper, as we concentrate on the vessel segmentation only.

Several methods for segmenting items of interest have been reported, using either rule-based or supervised methods for both fluorescein and non-mydriatic colour retinal images [14, 17, 19]. Mathematical morphology, which is a rule-based method, has previously revealed itself as a very useful digital image processing technique for detecting and counting microaneurysms in fluorescein and non-mydriatic camera images [4, 12, 18]. Wavelet transform theory has grown rapidly since the seminal work by Morlet and Grossman, finding applications in many realms (e.g. [9]). The wavelets space-scale analysis capability can be used to decompose vessel structures into differently scaled Morlet wavelets, so as to segment them from the retinal fundus.

The recognition of images, or parts of images as possessing pathologies, has responded well to automated classification techniques. Here the key is to determine some relationship between a set of input vectors that represent stimuli, and a corresponding set of values on a nominal scale that represent category or class. The relationship is obtained by applying an algorithm to training samples that are 2-tuples $(\mathbf{u}, z)$, consisting of an input vector $\mathbf{u}$ and a class label $z$. The learned relationship can then be applied to instances of $\mathbf{u}$ not included in the training set, in order to discover the corresponding class label $z$ [6]. This process, known as supervised classification, requires manually labelled images for training the model, and also requires suitable measures to form the vector $\mathbf{u}$. These measures can be derived from the previously discussed techniques, including mathematical morphology and the wavelet transform. After training, the model can then be used to classify previously unseen images. Alternatively, it is possible to classify individual pixels as either belonging to a vessel or to the background of the image. The classification technique can include Artificial Neural Networks or many others from the range of techniques available (e.g. [8, 14, 17, 19]).

# 3 Methods

In this work we assess the relative merits of several techniques for segmentation of blood vessels from colour retinal images. Twenty digital images were used from the Stare database [11]. This database also includes the opinions of two experts who had indicated the position of the vessels from colour images to establish two "gold standards" as separate images.

Our strategy was to use three methods for segmenting retinal blood vessels from directly digitized colour retinal images. The experimental procedure followed was to pre-process the images first to optimise the use of the wavelet transforms. The methods tested were:

1. Wavelet transform plus adaptive thresholding,
2. Wavelet transform plus supervised classifiers,
3. Wavelet transform plus pixel probabilities combined with adaptive thresholding.

In addition, we compared two training techniques: training on one or more complete images, then classifying the remaining images, and training on a window of the image then classifying the remainder of the same image.

Initially the methods were compared qualitatively, but the best of these methods were selected and compared numerically by plotting on a graph of true positive against false positive results from the classification. This graph resembles a free-response receiver operating characteristic (FROC) curve to aid the reader in its interpretation. True positives occur when the classifier labels a pixel as belonging to a vessel and the gold standard segmentation also labels the pixel as vessel.

In order to reduce the noise effects associated with the processing, the input image was pre-processed by a mean filter of size $5 \times 5$ pixels. Due to the circular shape of the non-mydriatic image boundary, neither the pixels outside the region-of-interest nor its boundary were considered, in order to avoid boundary effects. For our wavelet analysis we used the green channel of the RGB components of the colour image as it displayed the best vessels/background contrast.

### 3.1 Continuous Wavelet Transform Plus Adaptive Thresholding

Applying the continuous wavelet transform approach provides several benefits but resulted in some loss of detail as the scale parameter was fixed. We therefore adopted a pixel thresholding approach that represented each pixel by a feature vector including colour information, measurements at different scales taken from the continuous wavelet (Morlet) transform and the Gaussian Gradient, as well as from mean filtering applied to the green channel. The resulting feature space was used to provide an adaptive local threshold to assign each pixel as either a vessel-pixel or a non-vessel pixel.

The real plane $R \times R$ is denoted as $R^2$, and vectors are represented as bold letters, e.g. $\mathbf{x}, \mathbf{b} \in R^2$. Let $f \in L^2$ be an image represented as a square integrable (i.e. finite energy) function defined over $R^2$ [2]. The *continuous wavelet transform* (CWT) is defined as:

$$T_\psi(\mathbf{b}, \theta, a)(\mathbf{x}) = C_\psi^{-1/2} \frac{1}{a} \int \psi^*(a^{-1} r_{-\theta}(\mathbf{x} - \mathbf{b})) f(\mathbf{x}) d^2\mathbf{x} \tag{1}$$

where $C_\psi$, $\psi$, $\mathbf{b}$, $r_{-\theta}$, $\theta$ and $a$ denote the normalizing constant, the analysing wavelet, the displacement vector, the rotation operator, the rotation angle and the dilation parameter, respectively ($\psi^*$ denotes the complex conjugate). The double integral is taken over $R^2$ with respect to vector variable $\mathbf{x}$, being denoted by $d^2\mathbf{x}$. The Morlet wavelet is directional (in the sense of being effective in selecting orientations) and capable of fine tuning specific frequencies. These latter capabilities are especially important in filtering out the background noise, and comprise the advantages of the Morlet wavelet with respect to other standard filters such as the Gaussian and its derivatives. The 2D Morlet wavelet is defined as:

$$\psi_M(\mathbf{x}) = \exp(j\mathbf{k}_0 \cdot \mathbf{x}) \exp\left(-\frac{1}{2}|A\mathbf{x}|^2\right) \tag{2}$$

where $j = \sqrt{-1}$ and $A = \text{diag}\lfloor \epsilon^{-1/2}, 1 \rfloor, \epsilon \geq 1$ is a $2 \times 2$ array that defines the anisotropy of the filter, i.e. its elongation in some direction. In the Morlet equation (2), which is a complex exponential multiplying a $2D$ Gaussian, $\mathbf{k}_0$ is a vector that defines the frequency of the complex exponential. Using the Morlet transform to segment the blood vessels, the scale parameter is held constant and the transform is calculated for a set of orientations $\theta = 0, 10, 20, 30, ..., 180$. The $\epsilon$ parameter has been set as 4 in order to make the filter elongated and $\mathbf{k}_0 = [0, 2]$, i.e. a low frequency complex exponential with few significant oscillations. The transform maximum response (in modulus) from all orientations for each position, $\mathbf{b}$, is then taken, emphasizing the blood vessels and filtering out most of the noise. The blood vessels can then be detected from this representation.

### 3.2 Feature Extraction

The pixel feature space was formed by Morlet wavelet responses (taken at different scales and elongations), Gaussian Gradient responses (taken at different scales) and colour information, which determine each pixel's colour. This resulted in a computationally demanding high dimensional feature space. At the same time, Morlet responses taken at close scales are highly correlated, as are the Gaussian Gradient responses for similar scales. Therefore we used a feature extraction approach to obtain a lower dimensional feature space, while trying to preserve structure important for discrimination. Feature extraction was performed by a linear mapping provided by nonparametric discriminant analysis [7]. Nonparametric discriminant analysis consists of building two matrices. The first is a nonparametric between-class scatter matrix, constructed using k-nearest neighbour techniques, which defines the directions of class separability. The second is the within-class scatter matrix, which shows the scatter of samples around their mean class vectors. These matrices were built based on the labelled training samples. The two matrices are then used to find a projection (given by a linear mapping) that maximizes class separability while minimizing the within-class scatter in the projected feature space.

During the adaptive thresholding process, the dimensional nature of the features forming the feature space might give rise to errors. Since the feature space elements may be considered as random variables, we applied a normal transformation in order to obtain a new relative random variable, redefined in a dimensionless manner. The normal transformation is defined as:

$$\hat{X}_j = \frac{X_j - \mu_j}{\sigma_j} \tag{3}$$

where $X_j$ is the $j$th feature assumed by each pixel, $\mu_j$ is the average value of the $j$th feature and $\sigma_j$ is the associated standard deviation.

### 3.3 Supervised Classification

In methods 2 and 3, supervised classification was applied to obtain the final segmentation, with the pixel classes defined as C1 = vessel-pixels and

C2 = non-vessel pixels, using the Bayesian classifier consisting of a mixture of Gaussians [20]. In order to obtain the training set, retinal fundus images were manually segmented, thus allowing the creation of a labelled training set into two classes C1 and C2 (i.e. vessels and non-vessels). In this work, the hand-drawn vascular tree provided by the ophthalmologist was used - our training pattern - to obtain a feature space. Two different strategies for deriving the training set were applied:

1. Some images were completely segmented by an expert and a random sub-set of their pixels was used to train the classifier.
2. Only a small portion (window) of a sample image was manually segmented. The labelled pixels were then used to train the classifier, which was applied to the same image in order to complete its segmentation.

This second strategy was devised so that a semi-automated fundus segmentation software can be developed, in which the operator only has to draw a small portion of the vessels over the input image or simply click on several pixels associated with the vessels. The remaining image is then segmented based on this partial training set without the need of tuning any additional parameters. This approach requires a small effort from the operator, which is compensated for by the fact that image peculiarities (e.g. due to camera model and settings) are directly incorporated by the classifier. Note that this method should be repeated for every new image.

### 3.4 Post-processing

The output produced by the classifier leads to a binary image where each pixel is labelled as vessel or non-vessel. Some misclassified pixels appeared as undesirable noise in the classified image. In addition, for some vessels, only their boundaries were classified, so that it was necessary to perform post-processing by using morphological tools to obtain the final desired segmentation. Finally, to optimize the vessel contours, morphological operations have been applied, beginning by area open to eliminate small noisy components. The vessels were completely filled by morphological dilation and area close [3].

## 4 Results

In order to compare the results of these methods, we provide for comparison an example of the application of the wavelet transform to non-mydriatic images [13]. Figure 1(a) shows a typical image of the retinal fundus with the optic disc on the right hand side and the blood vessels that course throughout the image. Figure 1(b) shows the result of image segmentation using the Morlet wavelet transform with global thresholding. The latter shows the difficulty in obtaining a clear segmentation. Background noise and variable grey levels across the image introduce artifacts. In particular, this method did not remove all parts of the optic disc and was very susceptible to hue variation that resulted in areas of over sensitivity and under sensitivity in the same image.

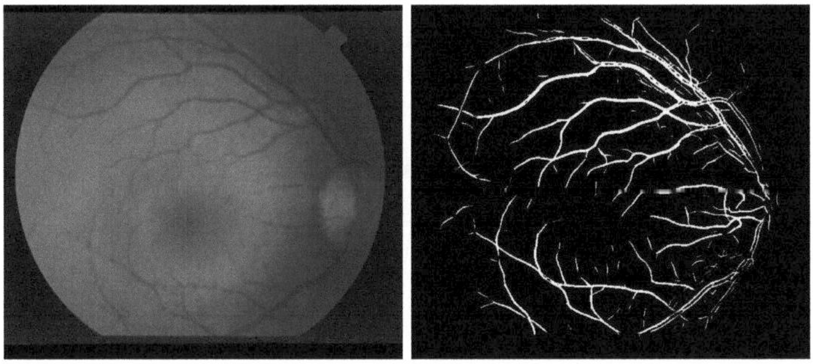

**Fig. 1.** Wavelet transform in blood vessel segmentation: a) original retinal image; b) example of segmentation of blood vessels using the Morlet wavelet transform with a global threshold

In method 1 we applied the wavelet transform plus adaptive thresholding to colour non-mydriatic camera images. Figure 2(a) shows a typical grey scale representation of a colour image obtained from the digital camera. The optic disc is noticeable as a light grey area on the left hand side with blood vessels emanating from it. Notice the variable brightness across the image, and especially the presence of the optic disc, which can introduce artifacts during the image processing. Figure 2(b) shows the same image after application of the Morlet wavelet transform and thresholding. This is much more successful than using global thresholding, as in Figure 1. The optic disc has been successfully removed, but artifacts remain. In particular, notice the extra vessels apparent at the bottom of Figure 2(b) at approximately 5 o'clock. Many disconnected segments also remain, and some smaller vessels clearly visible in (a) have not been detected in (b).

For a more sophisticated approach to dealing with the image variations in hue of background and blood vessels, we applied a supervised learning algorithm.

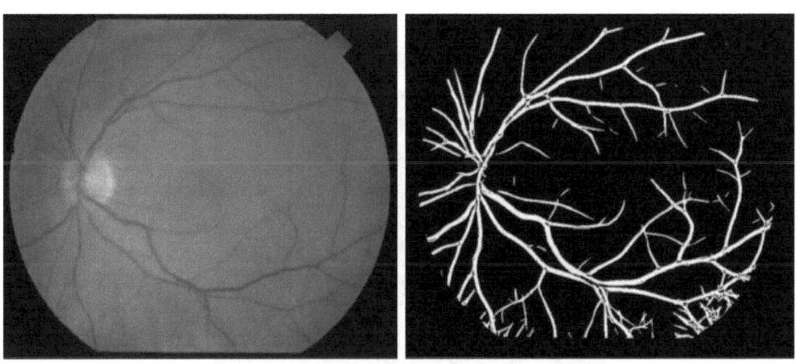

**Fig. 2.** Segmentation of non-mydriatic colour images for method 1: a) grey- scale image of original retinal fundus; b) segmentation of retinal blood vessels using wavelet transform and adaptive thresholding

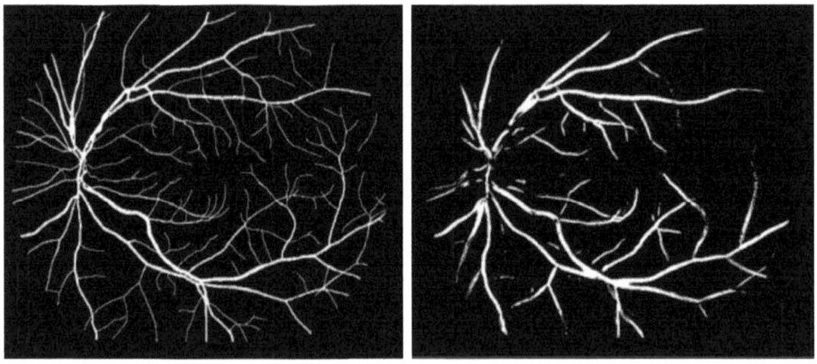

**Fig. 3.** Results from method 2, wavelet transform plus pixel classification: (a) an example of the training set provided by the experts; (b) an example of a segmented image obtained using the total vessel pattern as a training set

The classifier was first trained using all pixels from entire images. All pixels were labelled by the experts, as shown in Figure 3(a). The trained classifier was then used to segment other images. In Figure 3(b) we show the result of supervised classification, where the classifier has been trained on four other images, and then used to segment the image of Figure 2(a). Comparing this with Figure 2(b) the improvement is obvious. Many of the artifacts at the bottom (5 o'clock) of that image have now disappeared. However, many of the smaller vessels towards the centre of the image have not been detected, and there are still many disconnected vessel segments.

For method 3, we combined the wavelet transform with the supervised classification and mixed adaptive thresholding. In this case, instead of using the simplified approach of Leandro et al. the thresholding procedure was applied to the pixel probability of being vessel as estimated by the supervised classifier approach [13]. This led to the results shown in Figure 4. Here many of the smaller vessels are now visible, and there are far fewer disconnected vessel segments.

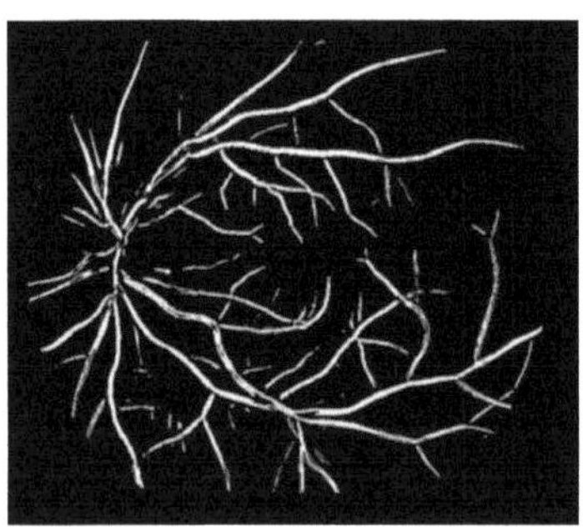

**Fig. 4.** The same image after adaptive thresholding on the probability of each pixel being part of a vessel

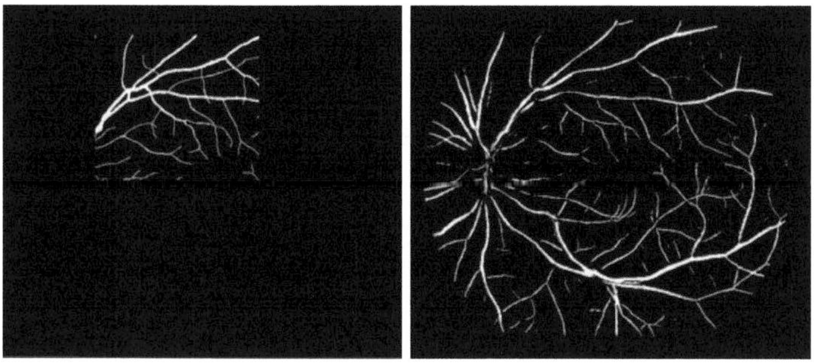

**Fig. 5.** Segmented image (b) obtained using only a portion of the vessel pattern as a training set (a)

A variation of the pixel classification is to train the classifier with a window of the image, then use it to segment the remainder of the image. This should provide more accurate classification, as it corrects for different image parameters. Figure 5(a) shows the window containing the training data for method 2. This represents a portion of the retinal vessels as identified by the expert. Figure 5(b) shows the result of the segmentation when only using a part of the figure as a training set. The number of small vessels detected has increased, and the segmentation is of superior quality. Compare this with figure 3.

Finally, we applied the adaptive thresholding (method 3) to the vessel probability of each pixel of the window based classification. A typical result is shown in Figure 6. This represents the best result obtained so far, where most of the smaller vessels have been detected. The main problem with this approach is that it does not take the probability of being background into account.

It is clear from these results that methods 2 and 3, each using the supervised classifier approach, provide the best results. We now present quantitative results from these two methods in Figure 7. For method 2, (wavelet transform and supervised classifier) each source image resulted in a single binary output image

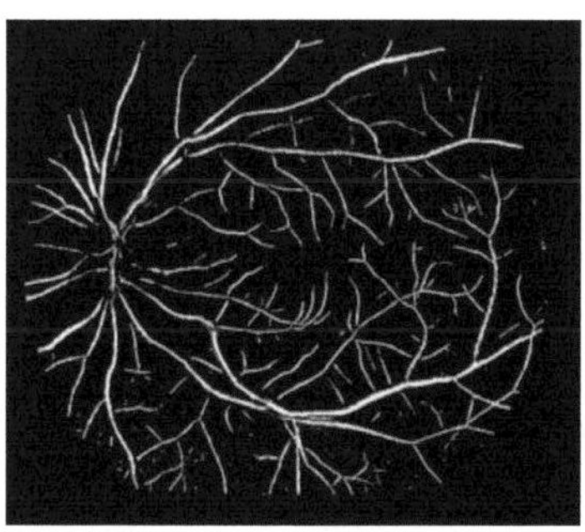

**Fig. 6.** Typical result of using the window method to train the classifier, followed by an adaptive thresholding process

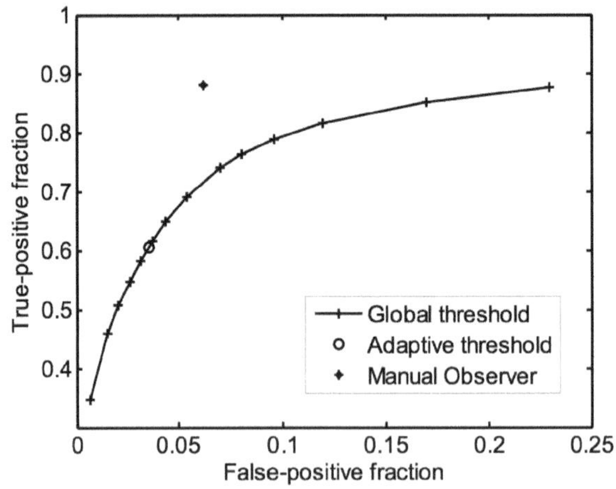


**Fig. 7.** Numerical results from methods 2 and 3, shown in the same form as an ROC graph. Method 2 uses an adaptive threshold, so a single point was obtained, being the average of the values obtained from the 20 images. Method 3 uses a global threshold, so many points were generated as the threshold was varied. Each point is the average of the results from the 20 images. The cross indicates the evaluation of one ophthalmologist.

with pixels either marked as 'true' (a vessel pixel) or marked "false" (not a vessel pixel). Each output image produced the single point on the graph in figure 7. The average from the 20 images processed is shown as a circle marked "Adaptive threshold" in the legend.

Method 3 (wavelet transform and adaptive threshold) resulted in 20 grey-scale images, where the brighter the pixel the more likely it landed in the vessel class. A global threshold was applied to each image to generate a point on the graph. The threshold was varied from high (poor sensitivity) to low (good sensitivity but too many false positives). The average values taken from the 20 images produced a number of points tracing out the curve (appearing in the legend as "Global threshold").

## 5 Discussion

We have demonstrated some new techniques for the automated processing of non-mydriatic images in the study of diabetic retinopathy that can certainly be extended to other contexts in pattern recognition. The results we have obtained so far suggest that pixel classification, in conjunction with wavelet transform and adaptive thresholding, can provide noise-robust vessel segmentation. The approach reported here improved on previous results by reducing the level of interaction required with the segmentation program, providing a useful tool for non-specialists such as community health workers in assessing fundus complications associated with diabetes [1, 3, 5, 10, 22]. Wavelets are especially suitable for detecting singularities (e.g. edges) in signals, extracting instantaneous frequencies, and performing fractal and multifractal analysis [1, 10]. Applying the wavelet transform allows noise filtering and blood vessel enhancement in a single step. Our results indicate that for the same false-positive fraction, the supervised learning with adaptive thresholding obtained a greater than 75% sensitivity compared to the ophthalmologist with approximately 90% (Figure 7. Although these methods are targeted at segmentation in retinal blood vessels, there is no reason why they may not be applied in other areas, especially in medical imaging, where it is necessary to extract intricate branching patterns from images with a noisy background.

## Acknowledgments

RMC and JS are grateful to FAPESP (Research Support Foundation of the State of São Paulo, Brazil) and to CNPq (Brazil's National Council for Scientific and Technological Development). HJ was in receipt of grants from CSU and Australian Diabetes Association. The authors also wish to acknowledge the contribution of Alan Luckie and Tien Wong for their expert advice on diabetic retinopathy and arteriolar narrowing.

## References

1. Antoine, J.P., Barache, D., Cesar Jr., R.M., da Costa, L.: Shape characterization with the wavelet transform. Signal Processing 62(3), 265–290 (1997)
2. Arnéodo, A., Decoster, N., Roux, S.G.: A wavelet-based method for multifractal image analysis. I. Methodology and test applications on isotropic and anisotropic random rough surfaces. The European Physical Journal B 15, 567–600 (2000)
3. Cesar Jr., R.M., Jelinek, H.F.: Segmentation of retinal fundus vasculature in non-mydriatic camera images using wavelets. In: Suri, J.S., Laxminarayan, S. (eds.) Angiography and plaque imaging, pp. 193–224. CRC Press, London (2003)
4. Cree, M., Luckie, M., Jelinek, H.F., Cesar, R., Leandro, J., McQuellin, C., Mitchell, P.: Identification and follow-up of diabetic retinopathy in rural health in Australia: an automated screening model. In: AVRO, Fort Lauderdale, USA 5245/B5569 (2004)
5. da Costa, L.F.: On neural shape and function. In: Proceedings of the World Congress on Neuroinformatics: ARGESIM / ASIM- Verlag Vienna, pp. 397–411 (2001)
6. Dietterich, T.G., Bakiri, G.: Solving Multiclass Learning Problems Via Error-Correcting Output Codes. Journal of Artificial Intelligence Research 2, 263–286 (1995)
7. Fukunaga, K.: Introduction to statistical pattern recognition, 2nd edn. Academic Press, Boston (1990)
8. Gardner, G.G., Keating, D., Williamson, T.H., Elliot, A.T.: Automatic detection of diabetic retinopathy using an artificial neural network: a screening tool. British Journal of Ophthalmology 80, 940–944 (1996)
9. Goupillaud, P., Grossmann, A., Morlet, J.: Cycle-octave and related transform in seismic signal analysis. Geoexploration 23, 85–102 (1984)

10. Grossmann, A.: Wavelet Transforms and Edge Detection. In: Albeverio, S., et al. (eds.) Stochastic Processes in Physics and Engineering. Reidel Publishing Company, Dordrecht (1988)
11. Hoover, A., Kouznetsova, V., Goldbaum, M.: Locating Blood Vessels in Retinal Images by Piecewise Threshold Probing of a Matched Filter Response. IEEE Transactions on Medical Imaging 19, 203–210 (2000)
12. Jelinek, H.F., Cree, M.J., Worsley, D., Luckie, A., Nixon, P.: An Automated Microaneurysm Detector as a Tool for Identification of Diabetic Retinopathy in Rural Optometric Practice. Clinical and Experimental Optometry 89(5), 299–305 (2006)
13. Leandro, J.J.G., Cesar Jr., R.M., Jelinek, H.F.: Blood vessels segmentation in retina: preliminary assessment of the mathematical morphology and of the wavelet transform techniques. In: Proceedings of SIBGRAPI 2001, Floriaopolis - SC, pp. 84–90. IEEE Computer Society Press, Los Alamitos (2001)
14. Leandro, J.J.G., Soares, J.V.B., Cesar Jr., R.M., Jelinek, H.F.: Blood vessel segmentation of non-mydriatic images using wavelets and statistical classifiers. In: Proceedings of the Brazilian Conference on Computer Graphics, Image Processing and Vision (Sibgrapi 2003), Sao Paulo, Brazil, pp. 262–269. IEEE Computer Society Press, Los Alamitos (2003)
15. McQuellin, C.P., Jelinek, H.F., Joss, G.: Characterisation of fluorescein angiograms of retinal fundus using mathematical morphology: a pilot study. In: Proceedings of the 5th International Conference on Ophthalmic Photography, Adelaide, p. 83 (2002)
16. Silink, M.: The diabetes epidemic: The case for a resolution on diabetes. Diabetic Endocrine Journal 34(suppl. 1), 3–4 (2006)
17. Sinthanayothin, C., Boyce, J., Williamson, C.T.: Automated localisation of the optic disc, fovea and retinal blood vessels from digital colour fundus images. British Journal of Ophthalmology 83(8), 902–912 (1999)
18. Spencer, T., Olson, J.A., McHardy, K., Sharp, P.F., Forrester, J.V.: An Image-Processing Strategy for the Segmentation and Quantification of Microaneurysms in Fluorescein Angiograms of the Ocular Fundus. Comput. Biomed. Res. 29, 284–302 (1996)
19. Staal, J.J., Abramoff, M.D., Niemeijer, M.A., Viergever, B., van Ginneken, B.: Ridge-based vessel segmentation in color images of the retina. IEEE Transactions on Medical Imaging 23(4), 501–509 (2004)
20. Theodoridis, S.: Pattern Recognition. Academic Press, Baltimore (1999)
21. Yogesan, K., Constable, I.J., Barry, C.J., Eikelboom, R.H., Tay-Kearney, M.L.: Telemedicine screening of diabetic retinopathy using a hand-held fundus camera. Telemedicine Journal 6(2), 219–223 (2000)
22. Zana, F., Klein, J.C.: Segmentation of vessel-like patterns using mathematical morphology and curvature evaluation. IEEE Transactions on Image Processing 10(7), 1010–1019 (2000)

# Multistage-Based Genetic Algorithm for Flexible Job-Shop Scheduling Problem

Mitsuo Gen[1,*], Jie Gao[2], and Lin Lin[1]

[1] Graduate School of Information, Production and Systems, Waseda University
gen@waseda.jp, linlin@aoni.waseda.jp
[2] School of Management, Xi'an Jiaotong University, Xi'an, 710049, China
calebgao@yahoo.com

**Abstract.** Flexible job shop scheduling problem (fJSP) is an extension of the traditional job shop scheduling problem (JSP), which provides a closer approximation to real scheduling problems. In this paper, a multistage-based genetic algorithm with bottleneck shifting is developed for the fJSP problem. The genetic algorithm uses two vectors to represent each solution candidate of the fJSP problem. Phenotype-based crossover and mutation operators are proposed to adapt to the special chromosome structures and the characteristics of the problem. The bottleneck shifting works over two kinds of effective neighborhood, which use interchange of operation sequences and assignment of new machines for operations on the critical path. In order to strengthen the search ability, the neighborhood structure can be adjusted dynamically in the local search procedure. The performance of the proposed method is validated by numerical experiments on three representative problems.

**Keywords:** Flexible job shop scheduling problem; Multistage-based genetic algorithms; Bottleneck shifting; Neighbourhood structure.

## 1 Introduction

In the job shop scheduling problem (JSP), there are $n$ jobs that must be processed on a group of $m$ machines. Each job $i$ consists of a sequence of $m$ operations ($o_{i1}$, $o_{i2, \ldots,} o_{im}$), where $o_{ik}$ (the $k$-th operation of job $i$) must be processed without interruption on a predefined machine $m_{ik}$ for $p_{ik}$ time units. The operations $o_{i1}$, $o_{i2, \ldots,} o_{im}$ must be processed one after another in the given order and each machine can process at most one operation at a time.

Flexible job shop scheduling problem (fJSP) is a generalization of the job shop and the parallel machine environment, which provides a closer approximation to a wide range of real manufacturing systems. In a flexible job shop, each job $i$ consists of a sequence of $n_i$ operations ($o_{i1}$, $o_{i2}$, …, $o_{in_i}$). The fJSP extends JSP by allowing an operation $o_{ik}$ to be executed by one machine out of a set $A_{ik}$ of given machines. The processing time of operation $o_{ik}$ on machine $j$ is $p_{ikj}$>0. The fJSP problem is to choose for each operation $o_{ik}$ a machine $M(o_{ik}) \in A_{ik}$ and a starting time $s_{ik}$ at which the operation must be performed.

Bruker and Schlie were among the first to address the fJSP problem [1]. They developed a polynomial algorithm for solving the flexible job shop scheduling problem with two jobs. Chambers developed a tabu search algorithm to solve the problem

M. Gen et al.: Intelligent and Evolutionary Systems, SCI 187, pp. 183–196.
springerlink.com 

[2]. Mastrolilli and Gambardella proposed two neighborhood functions for the fJSP problem [3]. They proposed a tabu search procedure and provided an extensive computational study on 178 fJSP problems and 43 JSP problems. Their approach found 120 new better upper bounds and 77 optimal solutions over the 178 fJSP benchmark problems and it was outperformed in only one problem instance.

Yang presented a new genetic algorithm (GA)-based discrete dynamic programming approach [4]. Kacem *et al.* proposed the approach by localization to solve the resource assignment problem, and an evolutionary approach controlled by the assignment model for the fJSP problem [5]. Wu and Weng considered the problem with job earliness and tardiness objectives, and proposed a multiagent scheduling method [6]. Xia and Wu treated this problem with a hybrid of particle swarm optimization and simulated annealing as a local search algorithm [7]. Zhang and Gen proposed a multistage operation-based genetic algorithm to deal with the fJSP problem from a point view of dynamic programming [8].

In this paper, a hybrid genetic algorithm (hGA) is employed to solve the fJSP problem. The genetic algorithm uses two representations to adapt to the nature of this problem. One representation is used in initialization and mutation, and the other is used for crossover operation. In order to strengthen the search ability, bottleneck shifting serves as a local search method under the framework of GA, which only investigates the neighbor solutions possible to improve the initial solution.

We formulate the fJSP problem in Section 2. Section 3 presents the details of the genetic algorithm. The bottleneck shifting method is presented in Section 5. In Section 5, we present computational study on several well-known fJSP benchmark problems and compare our results with the results obtained by previous approaches. Some final concluding remarks are given in Section 6.

## 2 Mathematical Formulation

The flexible job shop scheduling problem is as follows: $n$ jobs are to be scheduled on $m$ machines. Each job $i$ represents $n_i$ ordered operations. The execution of each operation $k$ of job $i$ (noted as $o_{ik}$) requires one machine $j$ selected from a set of available machines called $A_{ik}$, and will occupy that machine for $t_{ikj}$ time units until the operation is completed. The fJSP problem is to assign operations on machines and to sequence operations assigned on each machine, subject to the constraints that: a) The operation sequence for each job is prescribed, b) Each machine can process only one operation at a time. In this study, we manage to minimize the following three criteria:

- Makespan ($c_M$) of the jobs;
- Maximal machine workload ($w_M$), i.e., the maximum working time spent at any machine;
- Total workload ($w_T$), which represents the total working time over all machines.

The notation used in this paper is summarized in the following:

- Indices

  $i, h$: index of jobs, $i, h = 1, 2, \ldots, n$;

  $j$: index of machines, $j = 1, 2, \ldots, m$;

  $k, g$: index of operation sequences, $k, g = 1, 2, \ldots, n_i$

- Parameters
  $n$: total number of jobs;
  $m$: total number of machines;
  $n_i$: total number of operations of job $i$;
  $o_{ik}$: the $k$-th operation of job $i$;
  $A_{ik}$: the set of available machines for the operation $o_{ik}$;
  $t_{ikj}$: processing time of the operation $o_{ik}$ on machine $j$
- Decision variables

$$x_{ikj} = \begin{cases} 1, & \text{if machine } j \text{ is selected for the operation } o_{ik} \\ 0, & \text{otherwise} \end{cases}$$

$c_{ik}$ : completion time of the operation $o_{ik}$

The fJSP model is then given as follows:

$$\min \quad c_{\mathrm{M}} = \max_{1 \le i \le n} \{ c_{in_i} \} \tag{1}$$

$$\min \quad w_{\mathrm{M}} = \max_{1 \le j \le m} \left\{ \sum_{i=1}^{n} \sum_{k=1}^{n_i} t_{ikj} x_{ikj} \right\} \tag{2}$$

$$\min \quad w_{\mathrm{T}} = \sum_{i=1}^{n} \sum_{k=1}^{n_i} \sum_{j=1}^{m} t_{ikj} x_{ikj} \tag{3}$$

$$\text{s.t.} \quad c_{ik} - c_{i(k-1)} \ge t_{ikj} x_{ikj}, \ k = 2, \cdots, n_i; \ \forall \ i, j \tag{4}$$

$$[(c_{hg} - c_{ik} - t_{hgj}) x_{hgj} x_{ikj} \ge 0] \vee [(c_{ik} - c_{hg} - t_{ikj}) x_{hgj} x_{ikj} \ge 0], \ \forall \ (i,k), (h,g), j \tag{5}$$

$$\sum_{j \in A_{ik}} x_{ikj} = 1, \ \forall \ k, i \tag{6}$$

$$x_{ikj} \in \{0,1\}, \ \ \forall \ j, k, i \tag{7}$$

$$c_{ik} \ge 0, \ \forall \ k, i \tag{8}$$

Inequality (4) describes the operation precedence constraints. Inequality (5) is a disjunctive constraint, where one or the other constraint must be observed. It represents that the operation $o_{hg}$ should be not be started before the completion of operation $o_{ik}$, or that the operation $o_{hg}$ must be completed before the starting of operation $o_{ik}$ if they are assigned on the same machine $j$. Shortly, the execution of operation $o_{ik}$ cannot be overlapped in time with the execution of operation $o_{hg}$. Equation (6) states that one machine must be selected from a set of available machines for each operation.

# 3 Genetic Approach for SPR Problem

## 3.1 Genetic Representation

The GA's structure and parameter setting affect its performance. However, the primary determinants of a GA's success or failure are the coding by which its genotypes represent candidate solutions and the interaction of the coding with the GA's recombination and mutation operators.

As mentioned above, the fJSP problem is a combination of machine assignment and operation sequencing decisions. A solution can be described by the assignment of operations to machines and the processing sequences of operations on the machines. In this paper, the chromosome is therefore composed of two parts: a) Machine assignment vector (hereafter called $v_1$); b) Operation sequence vector (hereafter called $v_2$).

Consider a flexible job shop scheduling problem with four machines and four jobs, where each job requires four operations. It is rather easy to represent the machine assignment in one row. In each machine assignment vector $v_1$, $v_1(r)$ represents the machine selected for the operation indicated at locus $r$ (hereafter, we call it operation $r$ for shortness). An example of the machine assignment vector is shown in Fig. 1.

| Locus: $(r)$ | 1 | 2 | 3 | 4 | 5 | 6 | 7 | 8 | 9 | 10 | 11 | 12 | 13 | 14 | 15 | 16 |
|---|---|---|---|---|---|---|---|---|---|---|---|---|---|---|---|---|
| Operation Indicated | $o_{1,1}$ | $o_{1,2}$ | $o_{1,3}$ | $o_{1,4}$ | $o_{2,1}$ | $o_{2,2}$ | $o_{2,3}$ | $o_{2,4}$ | $o_{3,1}$ | $o_{3,2}$ | $o_{3,3}$ | $o_{3,4}$ | $o_{4,1}$ | $o_{4,2}$ | $o_{4,3}$ | $o_{4,4}$ |
| Machine Assignment: $v_1(r)$ | 4 | 3 | 3 | 1 | 2 | 4 | 1 | 4 | 3 | 1 | 2 | 1 | 2 | 4 | 4 | 3 |

**Fig. 1.** Illustration of the machine assignment vector

Permutation representation is perhaps the most natural representation of operation sequences. Unfortunately because of the existence of precedence constraints, not all the permutations of the operations define feasible sequences. For job shop scheduling problem, Gen and his colleagues proposed an alternative: they name all operations for a job with the same symbol and then interpret them according to the order of occurrence in the sequence of a given chromosome [9][10]. Gen and Zhang also applied this representation to advanced scheduling problem [11]. The method can also be used to represent the operation sequences for the fJSP problem [12]. Each job $i$ appears in the operation sequence vector ($v_2$) exactly $n_i$ times to represent its $n_i$ ordered operations. For example, the operation sequence represented in Fig. 2 can be translated into a list of ordered operations below:

$$o_{2,1} \succ o_{4,1} \succ o_{3,1} \succ o_{1,1} \succ o_{4,2} \succ o_{1,2} \succ o_{4,3} \succ o_{3,2} \succ o_{2,2} \succ o_{1,3} \succ o_{3,3} \succ o_{1,4} \succ o_{2,3} \succ o_{4,4} \succ o_{3,4} \succ o_{2,4}.$$

| Locus: Priority $(s)$ | 1 | 2 | 3 | 4 | 5 | 6 | 7 | 8 | 9 | 10 | 11 | 12 | 13 | 14 | 15 | 16 |
|---|---|---|---|---|---|---|---|---|---|---|---|---|---|---|---|---|
| Operation Sequence: $v_2(s)$ | 2 | 4 | 3 | 1 | 4 | 1 | 4 | 3 | 2 | 1 | 3 | 1 | 2 | 4 | 3 | 2 |

**Fig. 2.** Illustration of the operation sequence vector

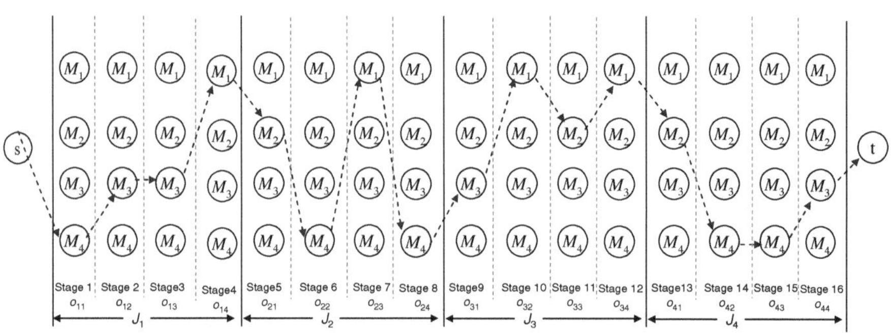


**Fig. 3.** Illustration of the two-vector representation

The main advantages of the two-vector representation are that each possible chromosome always represents a feasible solution candidate, and that the coding space is smaller than that of permutation representation. A simple example of the representation is shown in Fig. 3.

## 3.2 Priority-Based Decoding

In this paper we use priority-based decoding, where each operation searches the earliest available time interval for implementing on its assigned machine in the order represented by operation sequence vector. Given a time interval $[t_j^E, t_j^L]$ (beginning from $t_j^E$, and ending at $t_j^L$) on machine $j$, operation $o_{ik}$ will start as early as possible with its starting time as $\max\{t_j^E, c_{i(k-1)}\}$ (if $k \geq 2$) or $t_j^E$ (if $k=1$). Time interval $[t_j^E, t_j^L]$ is available for $o_{ik}$ if there is enough time span from the starting of $o_{ik}$ until the ending of the interval to complete it, i.e.,

$$\begin{cases} \max\left\{t_j^E, c_{i(k-1)}\right\} + t_{ikj} \leq t_j^L, & \text{if } k \geq 2; \\ t_j^E + t_{ikj} \leq t_j^L, & \text{if } k = 1. \end{cases} \tag{9}$$

The proposed priority-based decoding allocates each operation on its assigned machine one by one in the order represented by the operation sequence vector. When operation $o_{ik}$ is scheduled on machine $j$, the idle time intervals between operations that have already been scheduled on the machine are examined from left to right to find the earliest available one. If such an available interval exists, it is allocated there; otherwise, it is allocated at the end of machine $j$. The priority-based decoding method allows an operation to search the earliest available time interval on the machine.

Chromosomes are evaluated in phenotype space, while valuable information of the parental solutions is passed down to their children by means of manipulating chromosomes. In order to facilitate offspring to inherit the operation sequence information of their parents, it is necessary to unify the operation sequence in the chromosome with the sequence in the corresponding decoded schedule. The operation sequence in a chromosome is reordered according to the operation starting time of the decoded schedule before the chromosome involves crossover and mutation operations.

## 3.3 Phenotype-Based Crossover

In this study, the initial population is generated randomly in order to maintain the diversity of individuals. Starting from the initial population, genetic operations then evolve the population to converge to the optimal solution. Genetic operators mimic the process of heredity of genes to create new offspring at each generation.

In this study, crossover operators do not manipulate chromosomes in genotype space to generate offspring, but recombine schedules decoded from chromosomes to generate offspring in phenotype space. The decoded schedules can be expressed by two vectors: machine assignment vector ($v_1$) and operation starting time vector ($v_3$). The starting time vector describes the starting time of each operation, as shown in Fig. 4.

| Locus: $(r)$ | 1 | 2 | 3 | 4 | 5 | 6 | 7 | 8 | 9 | 10 | 11 | 12 | 13 | 14 | 15 | 16 |
|---|---|---|---|---|---|---|---|---|---|---|---|---|---|---|---|---|
| Operation Indicated | $o_{1,1}$ | $o_{1,2}$ | $o_{1,3}$ | $o_{1,4}$ | $o_{2,1}$ | $o_{2,2}$ | $o_{2,3}$ | $o_{2,4}$ | $o_{3,1}$ | $o_{3,2}$ | $o_{3,3}$ | $o_{3,4}$ | $o_{4,1}$ | $o_{4,2}$ | $o_{4,3}$ | $o_{4,4}$ |
| Machine Assignment: $v_1(r)$ | 4 | 3 | 3 | 1 | 2 | 4 | 1 | 4 | 3 | 1 | 2 | 1 | 2 | 4 | 4 | 3 |
| Starting Time: $v_3(r)$ | 0 | 21 | 33 | 51 | 0 | 94 | 112 | 136 | 0 | 21 | 45 | 69 | 16 | 32 | 62 | 94 |

**Fig. 4.** Representation of the decoded schedule

In the phenotype, an enhanced One-Cut-Point crossover is used to recombine two chromosomes. This type of crossover randomly selects one cut point at either the machine assignment or operation starting time vector, and then exchanges the down right parts of the two parents to generate offspring, as illustrated in Fig. 5.

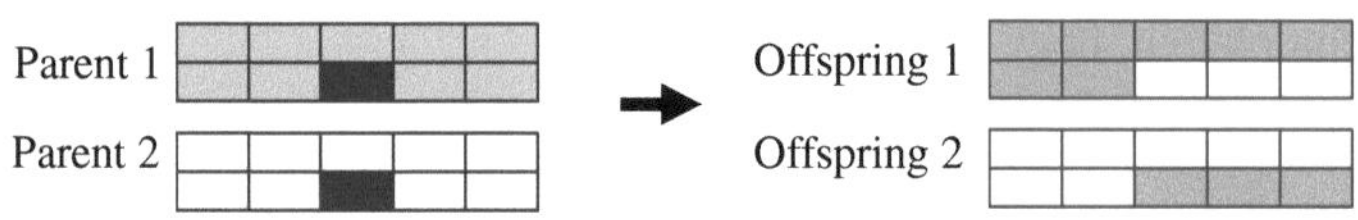


**Fig. 5.** Illustration of the enhanced one-cut-point crossover

### 3.4 Phenotype-Based Mutation

In this study, we use allele-based mutation. For machine assignment vectors, allele-based mutation randomly decides whether an allele should be selected for mutation in a certain probability. Then, a new available machine will be assigned for the operation indicated by the selected allele. For operation starting time vectors, two alleles are randomly selected, then the starting time in the two selected alleles are swapped.

The offspring schedules generated through crossover and mutation in the phenotype space cannot enter the population before they are encoded back to chromosomes. The machine assignment vector simply copies the machine assignment of the new born offspring schedule, and the operation sequence vector is generated by recording the job number of each operation in the order of their starting time in the offspring schedule from early to late. When two operations have the same starting time, the sequence between them is decided at random. The offspring operation sequences generated by order crossover are transformed back into the format of Gen *et al*'s representation by replacing each operation with its job number before they are released into the population. The order crossover does not ultimately generate any infeasible operation sequence vectors because Gen *et al*'s representation repairs them into feasible ones.

### 3.5 Fitness Function

The three considered objectives do not conflict with one another as seriously as in most other multiobjective optimization problems, because a small makespan ($c_M$) requires a small maximal workload ($w_M$) and a small maximal workload implies a small total workload ($w_T$). During evaluation, the fitness of a solution is calculated by synthesizing the three objectives into a weighted sum. We have to normalize the objective values on the three criteria before they are summed since they are of different scales. Let $c_M(l)$ be the makespan of the $l$-th chromosome. The scaled makespan ($c_M'(l)$) of a solution $l$ is as follows:

$$c_M'(l) = \begin{cases} \dfrac{c_M(l) - c_M^{\min}}{c_M^{\max} - c_M^{\min}}, \text{ if } c_M^{\max} \neq c_M^{\min}, \\ 0.5, \qquad \text{otherwise,} \end{cases} \quad \text{for all } l \tag{10}$$

where:

$$c_M^{\min} = \min_{1 \le l \le P} \{c_M(l)\};$$
$$c_M^{\max} = \max_{1 \le l \le P} \{c_M(l)\}$$

where $P$ is the total number of solution candidates to be evaluated in a generation. With the same method, we can scale maximal workload $w_M(l)$ and total workload $w_T(l)$ for each solution $l$. After scaling, the three objectives all take values from the range of [0, 1].

In order to guide the genetic and local search to the most promising area, makespan is given a very large weight since the other two objectives heavily depend on it. Additionally, it is typically the most important criterion in practical production environments. For the fJSP problem, a number of solutions with different maximal workloads or total workloads may have the same makespan. From this point of view, we firstly find the solutions with the minimum makespan, then minimize the maximal workload and the total workload in the presence of the minimum makespan. The fitness of a solution $l$ then is:

$$f(l) = \alpha_1 \cdot c_M'(l) + \alpha_2 \cdot w_M'(l) + \alpha_3 \cdot w_T'(l) \tag{11}$$

where $\alpha_1 > \alpha_2 > \alpha_3 > 0$ and $\alpha_1 + \alpha_2 + \alpha_3 = 1$.

### 3.6 Framework of the Algorithm

The overall structure of the multistage-based genetic algorithm can be illustrated as in Fig 6.

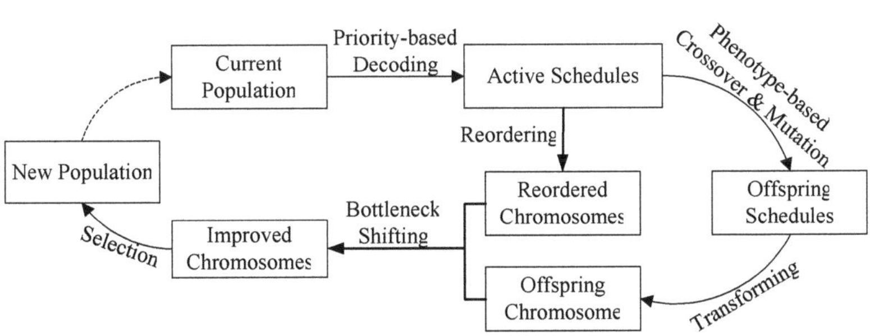


**Fig. 6.** Overall structure of the genetic algorithm

# 4 Bottleneck Shifting

## 4.1 Defining Neighborhood

A central problem of any local search procedure for combinatorial optimization problems is how to define the effective neighborhood around an initial solution. In this study, the effective neighborhood is based on the concept of critical path. To define neighborhood using critical path is not new for job shop scheduling problem and has been employed by many researchers [13 -16].

The feasible schedules of an fJSP problem can be represented with disjunctive graph $G = (N, A, E)$, with node set $N$, ordinary (conjunctive) arc set $A$, and disjunctive arc set $E$. The nodes of $G$ correspond to operations, the real arcs ($A$) to precedence relations, and the dashed arc ($E$) to pairs of operations to be performed on the same machine. For example, the following schedule of the 4×4 problem can be illustrated in the disjunctive graph shown in Fig. 7:

The Schedule={($o_{1,1}$, $M_4$: 0-16), ($o_{1,2}$, $M_3$: 21-33), ($o_{1,3}$, $M_3$:33-51 ), ($o_{1,4}$, $M_1$: 51-69), ($o_{2,1}$, $M_2$:0-16), ($o_{2,2}$, $M_4$:94-112), ($o_{2,3}$, $M_1$: 112-136), ($o_{2,4}$, $M_4$: 136-148), ($o_{3,1}$, $M_3$: 0-21), ($o_{3,2}$, $M_1$: 21-45), ($o_{3,3}$, $M_2$: 45-68), ($o_{3,4}$, $M_1$: 69-105), ($o_{4,1}$, $M_2$: 16-32), ($o_{4,2}$, $M_4$: 32-62), ($o_{4,3}$, $M_4$: 62-94), ($o_{4,4}$, $M_3$: 94-118)}.

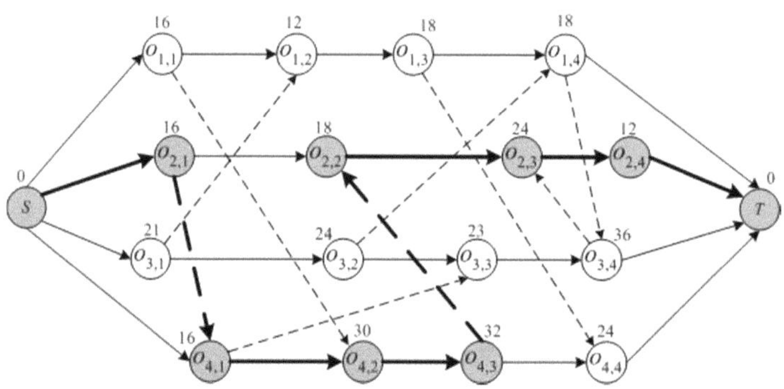


**Fig. 7.** Illustration of disjunctive graph

In Fig. 7, $S$ and $T$ are dummy starting and terminating nodes respectively. The number above each node represents the processing time of that operation. The critical path is the longest path in a graph. For an fJSP schedule, its makespan is equal to the length of the critical path in the corresponding disjunctive graph. The critical path is highlighted with broad-brush arcs in Fig. 7. Any operation on the critical path is called a critical operation. A critical operation cannot be delayed without increasing the makespan of the schedule.

The job predecessor $PJ(r)$ of an operation $r$ is the operation preceding $r$ in the operation sequence of the job that $r$ belongs to. The machine predecessor $PM(r)$ of an operation $r$ is the operation preceding $r$ in the operation sequence on the machine that $r$ is processed on. If an operation $r$ is critical, then at least one of $PJ(r)$ and $PM(r)$ must be critical, if they exist. In this study, if a job predecessor and a machine predecessor of a critical operation are both critical, then choose the predecessor (from these two alternatives) that appears first in the operation sequence.

A new schedule that is slightly different from the initial solution can be generated by changing the processing sequence of two adjacent operations performed on the same machine, *i.e.*, reversing the direction of the disjunctive arc that links the two operations. The neighborhood created in this way is named as type I here. Neighbor solutions can also be generated by assigning a different machine for one operation. This kind of neighborhood is named as type II.

The makespan of a schedule is defined by the length of its critical path, in other words, the makespan is no shorter than any possible path in the disjunctive graph. Hence, for a neighbor solution of type I, only when these two adjacent operations are on the critical path, the new solution is possible to be superior to the old one. Likewise, for a neighbor solution of type II, it cannot outperform the initial solution if the operation is not a critical one.

For the fJSP problem, we can only swap the operation sequence between a pair of operations that belong to different jobs. It is possible to decompose the critical path into a number of blocks, eachof which is a maximal sequence of adjacent critical operations that require the same machine. As a result, the possible swaps are further confined as follows:

- In each block, we only swap the last two and first two operations;
- For the first (last) block, we only swap the last (first) two operations in the block. In case where the first (last) block contains only two operations, these operations are swapped.
- If a block contains only one operation, then no swap is made.

Due to the strict restrictions above, possible swaps occur only on a few pairs of adjacent operations that belong to different jobs on the critical path. Neighbor solutions of type I are actually generated by implementing these possible swaps. Fig. 8 shows the critical path, critical blocks and the possible swaps in a schedule. The total number of the neighbors of type I ($N^I$) is less than the total number of critical operations ($N^C$) since some critical operations can not involve the possible swaps.

A neighbor solution of type II can be created by assigning a different machine $j \in A_{ik}$ for a critical operation $o_{ik}$. Let $n_l^{II}$ be the number of machines on which the $l$-th critical operation can be assigned. $n_l^{II}-1$ neighbors can be generated by assigning the

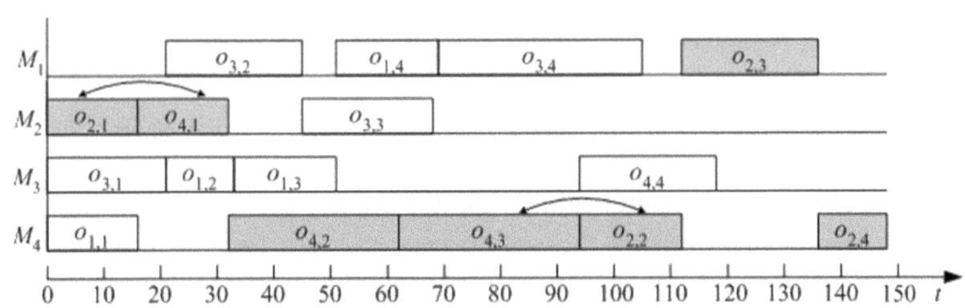


**Fig. 8.** Neighborhood of type I

operation on any of the other $n_l^{\mathrm{II}} - 1$ available machines. Hence, the total number of neighbors of type II ($N^{\mathrm{II}}$):

$$N^{\mathrm{II}} = \sum_{l=1}^{N^{\mathrm{C}}} n_l^{\mathrm{II}} - 1 \tag{12}$$

Since $N^{\mathrm{I}}$ is less than $N^{\mathrm{C}}$, $N^{\mathrm{II}}$ generally represents a much larger number than $N^{\mathrm{I}}$.

### 4.2 Local Search Transition Mechanism

During the local search, the original schedule will transit to a better neighbor solution, if it exists. This gives rise to a new problem: what is an improved solution. For the fJSP problem, there may be more than one critical path in a schedule, in which the makespan is determined by the length of the critical path. A solution with a smaller number of critical paths may provide more potential to find solutions with less makespan nearby because the makespan cannot be decreased without breaking all the current critical paths. An important problem of any local search method is how to guide to the most promising areas from an initial solution. In this study, a solution is taken to be an improved solution if it satisfies either of the two alternative requirements:

- An improved solution has a larger fitness value than the initial solution; or
- The improved solution has the same fitness value as the initial solution, yet it has less critical paths.

### 4.3 Adjust Neighborhood Structure

Let $N(i)$ denote the set of neighborhood of solution $i$. The enlarged two-pace neighborhood can be defined as the union of the neighborhood of each neighbor of the initial solution. Let $N^2(i)$ be the two-pace neighborhood of solution $i$, then,

$$N^2(i) = \cup_{j \in N(i)} N(j) \tag{13}$$

A larger neighborhood space size generally indicates a higher quality of the local optima because in each step of the local search, the best solution among a larger number of neighbor solutions is selected as the initial solution for the next local search iteration. On the other hand, a larger neighborhood space size would bring a greater computational load because more neighbor solutions have to be evaluated and compared. That is, each step of the local search will take longer time. Hence, the number of the local search iterations is decreased when the time spent on local search is limited. As a result, the deep search ability is not fully utilized.

In order to enhance the search ability of the local search without incurring too much computational load, during the search process over type II neighborhood, the local search procedure will implement over the enlarged two-pace neighborhood only when it reaches the local optimum of the one-pace neighborhood.

## 5 Experiments

In order to test the effectiveness and performance of the proposed hybrid genetic algorithm, three representative instances (represented by problem $n \times m$) were selected for simulation. The works by Kacem *et al.*[5][17], Xia and Wu [7], and Zhang and Gen [8] are among the most recent progresses made in the area of fJSP. Unfortunately, the simulation results are not included in their work. Hence, the results obtained by our method are compared with the results from [5] [7] and [17]. All the simulation experiments were performed with Delphi on Pentium 4 processor (2.6-GHz clock). The adopted parameters of the hGA are listed in table 1.

**Table 1.** Parameters of the hGA

| Parameters | Value | Parameters | Value |
|---|---|---|---|
| population size | 1500 | immigration mutation probability | 0.15 |
| maximal generation | 300 | $\alpha_1$ | 0.85 |
| order crossover probability | 0.3 | $\alpha_2$ | 0.10 |
| allele-based mutation probability | 0.10 | $\alpha_3$ | 0.05 |

### 5.1 Problem 8×8

This is an instance of partial flexibility. In the flexible job shop, there are 8 jobs with 27 operations to be performed on 8 machines. For more details about this problem, refer to [7]. Experimental simulations were run for 20 times. The 20 runs all converge to optimal solutions with the same objective values on the three considered criteria. One of the optimal solutions is shown in Fig. 9. This test instance seems to be oversimplified. It takes averagely 16.4 generations for the hGA to converge to the optimal solutions. The computation time averages at 5 minutes.

### 5.2 Problem 10×10

For this test instance, there are 10 jobs with 30 operations to be performed on 10 machines. For more details about this problem, refer to [7]. Experimental simulations were run for 20 times for this problem. The 20 runs all converge to optimal solutions with the same objective values. Averagely, the hGA takes 26.50 evolution generations and about 17 minutes to find the optimal solutions.

### 5.3 Problem 15×10

A larger-sized problem is chosen to test the performance of our hybrid genetic algorithm. This problem contains 15 jobs with 56 operations that have to be processed on 10 machines with total flexibility (for more details about this problem, refer to [7]).

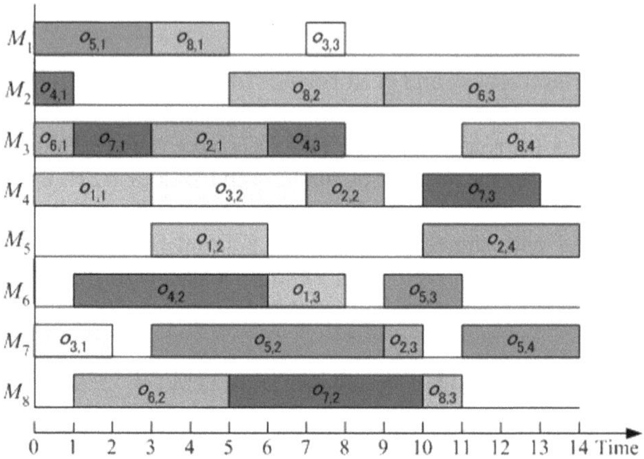


**Fig. 9.** Best solution 1 of problem 8×8 ($c_M$=14, $w_M$=12, $w_T$=77)

**Table 2.** Performance of the hGA for the three fJSP problems

| Problem | | Traditional GA | AL+CGA | | PSO+SA | | Proposed hGA |
|---|---|---|---|---|---|---|---|
| 8×8 | $c_M$ | 16 | 15 | 16 | 15 | 16 | 15 |
| | $w_M$ | | | | 12 | 13 | 12 |
| | $w_T$ | 77 | 79 | 75 | 75 | 73 | 75 |
| 10×10 | $c_M$ | 7 | 7 | | 7 | | 7 |
| | $w_M$ | 7 | 5 | | 6 | | 5 |
| | $w_T$ | 53 | 45 | | 44 | | 43 |
| 15×10 | $c_M$ | 23 | 24 | | 12 | | 11 |
| | $w_M$ | 11 | 11 | | 11 | | 11 |
| | $w_T$ | 95 | 91 | | 91 | | 91 |

Experimental simulations were run for 20 times for this problem. The 20 runs all converge to optimal solutions with the same fitness values. Not only different solutions with the same optimal fitness value but also distinct solutions with the same optimal objective values in the three considered criteria are found in one run of the experiment. Providing more than one good solution for decision-makers is a main advantage of genetic algorithms.

It takes about 97.75 generations to converge to the optimal solutions. It seems quite easy for our algorithm to find the optimal solutions even for such a large-sized fJSP problem. Yet, the time spent on finding the optimal solutions is rather long and averages at 135.47 minutes because the local search consumes large amount of computation time during the evolutionary process. In comparison with the scheduling horizon, the computation time is acceptable in the real world.

Table 2 gives the performance of the proposed method compared with other algorithms. "Approach by Localization" and "AL+CGA" are two algorithms by Kacem *et al.*[5] [17]. "PSO+SA" is the algorithm by Xia and Wu [7]. $c_M$, $w_M$, $w_T$ represent the makespan, the maximal machine workload and the total workload of the results for the three problems obtained by the approaches.

## 6 Conclusion

We have developed a new approach hybridizing genetic algorithm with bottleneck shifting to fully exploit the "global search ability" of genetic algorithm and "the local search ability" of bottleneck shifting for solving multiobjective flexible job shop scheduling problem. An innovative two-vector representation scheme is proposed and an effective decoding method is used to transform each chromosome into an active schedule. The initialization and mutation operations modify chromosomes of the representation. However, in order to enhance the heritability of crossover operation, chromosomes of the representation are transformed into the format of the two-vector permutation representation, and then an enhanced order crossover is proposed to implement recombination operation on the chromosomes of the two-vector permutation representation.

Two kinds of neighborhood are defined based on the concept of critical path for the fJSP problem. The two kinds of neighborhood are quite effective in that they only contain solutions that are likely to improve the initial solution. In the local search, the number of critical paths serves as one kind of intermediate objective besides the three original criteria in order to guide the local search to the most promising areas. The neighborhood structure can be dynamically adjusted during the local search process so that the quality of the local optima can be improved without incurring too much computational load.

Several well-known benchmark problems of different scales are solved by the proposed algorithm. The simulation results obtained in this study are compared with the results obtained by other algorithms. The results demonstrate the performance of the proposed algorithm.

**Acknowledgments.** The authors would like to say thanks to the two anonymous reviewers for their valuable comments. This work is partly supported by the Ministry of Education, Science and Culture, the Japanese Government: Grant-in-Aid for Scientific Research (No.19700071, No.20500143) and the National Natural Science Foundation of China (NSFC) under Grant No. 70433003.

## References

1. Bruker, P., Schlie, R.: Job-shop scheduling with multi-purpose machines. Computing 45, 369–375 (1990)
2. Chambers, J.B.: Classical and Flexible Job Shop Scheduling by Tabu Search. PhD thesis, University of Texas at Austin, Austin, U.S.A (1996)
3. Mastrolilli, M., Gambardella, L.M.: Effective neighborhood functions for the flexible job shop problem. J. Sched 3, 3–20 (2000)

4. Yang, J.-B.: GA-based discrete dynamic programming approach for scheduling in FMS environments. IEEE Trans. Systems, Man, and Cybernetics—Part B 31(5), 824–835 (2001)
5. Kacem, I., Hammadi, S., Borne, P.: Approach by localization and multiobjective evolutionary optimization for flexible job-shop scheduling problems. IEEE Trans. Systems, Man, and Cybernetics—Part C 32(1), 1–13 (2002)
6. Wu, Z., Weng, M.X.: Multiagent scheduling method with earliness and tardiness objectives in flexible job shops. IEEE Trans. System, Man, and Cybernetics—Part B 35(2), 293–301 (2005)
7. Xia, W., Wu, Z.: An effective hybrid optimization approach for muti-objective flexible job-shop scheduling problem. Computers & Industrial Engineering 48, 409–425 (2005)
8. Zhang, H., Gen, M.: Multistage-based genetic algorithm for flexible job-shop scheduling problem. Journal of Complexity International 11, 223–232 (2005)
9. Cheng, R., Gen, M., Tsujimura, Y.: A tutorial survey of job-shop scheduling problems using genetic algorithms-I. Representation. Computers & Industrial Engineering 30(4), 983–997 (1996)
10. Cheng, R., Gen, M., Tsujimura, Y.: A tutorial survey of job-shop scheduling problems using genetic algorithms, part II: hybrid genetic search strategies. Computers & Industrial Engineering 36(2), 343–364 (1999)
11. Gen, M., Zhang, H.: Effective Designing Chromosome for Optimizing Advanced Planning and Scheduling. In: Dagli, C.H., et al. (eds.) Intelligent Engineering Systems Through Artificial Neural Networks, vol. 16, pp. 61–66. ASME Press (2006)
12. Gao, J., Gen, M., Sun, L., Zhao, X.: A hybrid of genetic algorithm and bottleneck shifting for multiobjective flexible job shop scheduling problems. Computers & Industrial Engineering 53(1), 149–162 (2007)
13. Gen, M., Cheng, R.: Genetic Algorithms & Engineering Optimization. Wiley, New York (2000)
14. Adams, J., Balas, E., Zawack, D.: The shifting bottleneck procedure for job shop scheduling. Management Science 34(3), 391–401 (1998)
15. Balas, E., Vazacopoulos, A.: Guided local search with shifting bottleneck for job shop scheduling. Management Science 44(2), 262–275 (1998)
16. Goncalves, J.F., Mendes, J.J.M., Resende, M.G.C.: A hybrid genetic algorithm for the job shop scheduling problem. European Journal of Operational Research 167, 77–95 (2005)
17. Kacem, I., Hammadi, S., Borne, P.: Pareto-optimality approach for flexible job-shop scheduling problems: Hybridization of evolutionary algorithms and fuzzy logic. Mathematics and Computers in Simulation 60, 245–276 (2002)

# Implementation of Parallel Genetic Algorithms on Graphics Processing Units

Man Leung Wong[1] and Tien Tsin Wong[2]

[1] Department of Computing and Decision Sciences, Lingnan University, Tuen Mun, Hong Kong
mlwong@ln.edu.hk

[2] Department of Computer Science and Engineering, The Chinese University of Hong Kong, Shatin, Hong Kong
ttwong@acm.org

In this paper, we propose to parallelize a Hybrid Genetic Algorithm (HGA) on Graphics Processing Units (GPUs) which are available and installed on ubiquitous personal computers. HGA extends the classical genetic algorithm by incorporating the Cauchy mutation operator from evolutionary programming. In our parallel HGA, all steps except the random number generation procedure are performed in GPU and thus our parallel HGA can be executed effectively and efficiently. We suggest and develop the novel pseudo-deterministic selection method which is comparable to the traditional global selection approach with significant execution time performance advantages. We perform experiments to compare our parallel HGA with our previous parallel FEP (Fast Evolutionary programming) and demonstrate that the former is much more effective and efficient than the latter. The parallel and sequential implementations of HGA are compared in a number of experiments, it is observed that the former outperforms the latter significantly. The effectiveness and efficiency of the pseudo-deterministic selection method is also studied.

## 1 Introduction

Since Genetic Algorithms (GAs) were introduced in 1960s [1], several researchers have demonstrated that GAs are effective and robust in handling a wide range of difficult real-world problems such as feature selection [2], optimization [3], and data mining [4, 5, 6]. In general, GAs use selection, mutation, and crossover to generate new search points in a search space. An genetic algorithm starts with a set of individuals of the search space. This set forms a population of the algorithm. Usually, the initial population is generated randomly using a uniform distribution. On each iteration of the algorithm, each individual is evaluated using the fitness function and the termination function is invoked to determine whether the termination criteria have been satisfied. The algorithm terminates if acceptable solutions have been found or the computational resources have been spent. Otherwise, a number of individuals are selected and copies of them

M. Gen et al.: Intelligent and Evolutionary Systems, SCI 187, pp. 197–216.
springerlink.com 

replace individuals in the population that were not selected for reproduction so that the population size remains constant. Then, the individuals in the population are manipulated by applying different evolutionary operators such as mutation and crossover. Individuals from the previous population are called parents while those created by applying evolutionary operators to the parents are called offspring. The consecutive processes of selection, manipulation, and evaluation form a generation of the algorithm.

Although GAs are effective in solving many practical problems in science, engineering, and business domains, they may execute for a long time to find solutions for some huge problems, because several fitness evaluations must be performed. A promising approach to overcome this limitation is to parallelize these algorithms for parallel, distributed, and networked computers. However, these computers are relatively more difficult to use, manage, and maintain. Moreover, some people may not have access to this kind of computers.

Recently, more and more researchers suggest that the Graphics Processing Unit (GPU), which was originally designed to execute parallel operations for real-time 3D rendering, is a promising and convenient platform for performing general purpose parallel computation [7, 8, 9, 10, 11] because these GPUs are available in ubiquitous personal computers. Given the ease of use, maintenance, and management of personal computers, more people will be able to implement parallel algorithms to solve difficult and time-consuming problems encountered in real-world applications.

In [12, 13], we proposed to parallelize the Fast Evolutionary Programming (FEP) [14, 15, 16, 17] on GPU. Similar to Genetic Algorithms (GAs) [1, 3], Evolutionary Programming (EP) is a kind of population-based Evolutionary Algorithms (EAs) [16, 18]. One of the main differences between EP and GAs is that the former applies only the mutation operator to create new individuals, while GAs use the mutation and crossover operators to generate new offspring. In our parallel FEP, fitness value evaluation, mutation, and reproduction are executed in GPU. Since selection and replacement involve a global comparison procedure, they cannot be implemented efficiently in *Single-Instruction-Multiple-Data* (SIMD) based GPU. Consequently, selection and replacement are performed in CPU. The random numbers used by FEP are also generated by CPU because current GPU is not equipped with a random number generator. We compared our parallel FEP and an ordinary FEP on CPU. It was found that the speed-up factor of our parallel FEP ranges from 1.25 to 5.02, when the population size is large enough.

In this paper, we study a GPU implementation of a Hybrid Genetic Algorithm (HGA) that extends the classical genetic algorithm [3] by incorporating the Cauchy mutation operator from evolutionary programming [16, 14, 15]. All steps of HGA except the random number generation procedure are executed in GPU. Thus this parallel HGA is expected to be more effective and efficient than our previous parallel FEP.

In the following section, different parallel and distributed GAs will be described. GPU will be discussed in Section 3. We will present our parallel HGA in Sections 4 and 5. A number of experiments have been performed and the experiment results will be discussed in Section 6. We will give a conclusion and a description of our future work in the last section.

## 2 Parallel and Distributed Genetic Algorithms

For almost all practical applications of GAs, most computation time is consumed in evaluating the fitness value of each individual in the population since the genetic operators of GAs can be performed efficiently. Memory availability is another important problem of GAs because the population usually has a large number of individuals.

There is a relation between the difficulty of the problem to be solved and the size of the population. In order to solve substantial and real-world problems, a population size of thousands and a longer evolution process are usually required. A larger population and a longer evolution process imply more fitness evaluations must be conducted and more memory are required. In other words, a lot of computational resources are required to solve substantial and practical problems. Usually, this requirement cannot be fulfilled by normal workstations. Fortunately, these time-consuming fitness evaluations can be performed independently for each individual in the population and individuals in the population can be distributed among multiple computers.

GAs have a high degree of inherent parallelism which is one of the motivation of studies in this field. In natural populations, thousands or even millions of individuals exist in parallel and these individuals operates independently with a little cooperation and/or competition among them. This suggests a degree of parallelism that is directly proportional to the population size used in GAs. There are different ways of exploiting parallelism in GAs: master-slave models; improved-slave models; fine-grained models; island models; and hybrid models [19].

The most direct way to implement a parallel GA is to implement a global population in the master processor. The master sends each individual to a slave processor and let the slave to find the fitness value of the individual. After the fitness values of all individuals are obtained, the master processor selects some individuals from the population using some selection method, performs some genetic operations, and then creates a new population of offspring. The master sends each individual in the new population to a slave again and the above process is iterated until the termination criterion is satisfied.

Another direct way to implement a parallel GA is to implement a global population and use the tournament selection which approximates the behavior of ranking. Assume that the population size $N$ is even and there are more than $N/2$ processors. Firstly, $N/2$ slave processors are selected. A processor selected from the remaining processors maintains the global population and controls the overall evolution process and the $N/2$ slave processors. Each slave processor performs two independent $m$-ary tournaments. In each tournament, $m$ individuals are sampled randomly from the global population. These $m$ individuals are evaluated in the slave processor and the winner is kept. Since there are two tournaments, the two winners produced can be crossed in the slave processor to generate two offspring. The slave processor may perform further modifications to the offspring. The offspring are then sent back to the global population and the master processor proceeds to the next generation if all offspring are received from the $N/2$ slave processors.

Fine-grained GAs explore the computing power of massively parallel computers such as the Maspar. To explore the power of this kind of computers, one can assign one individual to each processor, and allow each individual to seek a mate close to it. A global random mating scheme is inappropriate because of the limitation of the communication abilities of these computers. Each processor can select probabilistically an individual in its neighborhood to mate with. The selection is based on the fitness proportionate selection, the ranking, the tournament selection, and other selection methods proposed in the literature. Only one offspring is produced and becomes the new resident at that processor. The common property of different massively parallel evolutionary algorithms is that selection and mating are typically restricted to a local neighborhood.

Island models can fully explore the computing power of course grain parallel computers. Assume that we have 20 high performance processors and have a population of 4000 individuals. We can divide the total population into 20 subpopulations (islands or demes) of 200 individuals each. Each processor can then execute a normal evolutionary algorithm on one of these subpopulations. Occasionally, the subpopulations would swap a few individuals. This migration allows subpopulations to share genetic material. Since there are 20 independent evolutionary searches occur concurrently, these searches will be different to a certain extent because the initial subpopulations will impose a certain sampling bias. Moreover, genetic drift will tend to drive these subpopulations in different directions. By employing migration, island models are able to exploit differences in the various subpopulations. These differences maintain genetic diversity of the whole population and thus can prevent the problem of premature convergence.

Hybrid models combine several parallelization approaches. The complexity of these models depends on the level of hybridization.

## 3 Graphics Processing Unit

In the last decade, the need from the multimedia and games industries for accelerating 3D rendering has driven several graphics hardware companies devoted to the development of high-performance parallel graphics accelerator. This resulted in the birth of GPU (Graphics Processing Unit), which handles the rendering requests using 3D graphics application programming interface (API). The whole pipeline consists of the transformation, texturing, illumination, and rasterization to the framebuffer. The need for cinematic rendering from the games industry further raised the need for programmability of the rendering process. Starting from the recent generation of GPUs launched in 2001 (including nVidia GeforceFX series and ATI Radeon 9800 and above), developers can write their own C-like programs, which are called *shaders*, on GPU. Due to the wide availability, programmability, and high-performance of these consumer-level GPUs, they are cost-effective for, not just game playing, but also scientific computing.

These shaders control two major modules of the rendering pipeline, namely vertex and fragment engines. As an illustration to the mechanism in GPU, we describe the rendering of a texture-mapped polygon. The user first defines the 3D position of each vertex through the API in graphics library (OpenGL or DirectX). It seems irrelevant to define 3D triangles for evolutionary computation. However, such declaration is necessary for satisfying the input format of the graphics pipeline. In our application, we simply define 2 triangles that cover the whole screen. The texture coordinate associating with each vertex is also defined at the same time. These texture coordinates are needed to define the correspondence of elements in textures (input/output data) and the pixels on the screen (shaders are executed on per-pixel basis). The defined vertices are then passed to the vertex engine for transformation (dummy in our case).

For each vertex, a vertex shader (user-defined program) is executed (Fig. 1). The shader program must be *Single-Instruction-Multiple-Data* (SIMD) in nature, *i.e.* the same set of operations has to be executed on different vertices. Then the polygon is then projected onto the 2D screen and rasterized (discretized) into many fragments (pixels) in the framebuffer as shown in Fig. 1. From now on, the two terminologies, pixel and fragment, are interchangeable throughout this paper. Next, the fragment engine takes place. For each pixel, a user-defined fragment shader is executed to process data associated with this pixel. Inside the shader, the input textures can be fetched for computation and results are output via the output textures. Again, the fragment shader must also be SIMD in nature.

One complete execution of the fragment shader is referred as one rendering pass. On a current GPU, there is a significant overhead for each rendering pass. The more rendering passes are needed, the slower the program is. Since fragment shaders are executed independently on each pixel, no information sharing is allowed among pixels. If the computation result of a pixel $A$ has to be used for computing an equation at pixel $B$, the computation result of $A$ must be written to an output texture first. This output texture has to be fed to the shader for computation in *next* rendering pass. Therefore, if the problem being tackled involves a chain of data dependency, more rendering passes are needed, and hence the speed-up is decreased.

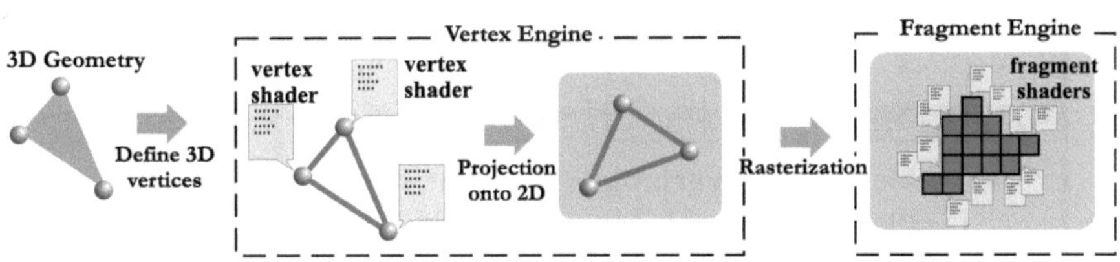


**Fig. 1.** The 3D rendering pipeline

The shading languages are high-level programming languages and closely resemble to C. Most mathematical functions available in C are supported by the shading language. Moreover, 32-bit floating point computation is supported on GPU. Hence, the GPU can be utilized for speeding up the time-consuming fitness evaluation in GA. Unfortunately, bit-wise operators are not well supported. Pseudo-random number generators that relying on bit-wise operations are not avaliable on current GPUs. Due to the SIMD architecture of GPU, certain limitations are imposed to the shading language. Data-dependent for-loop are not allowed because each shader may perform different number of iterations. Moreover, if-then-else construct is also not efficient, as the GPU will execute both true- and false-statements in order to comply with the SIMD design.

## 4 Data Organization

Suppose we have $\mu$ individuals and each contains $k$ variables. The most natural representation for an individual is an array. As GPU is tailored for parallel processing and optimized multi-channel texture fetching, all input data to GPU should be loaded in the form of *textures*. A texture is basically an image with each pixel composed of four components, $(r, g, b, \alpha)$. Each component can be represented as 32-bit floating point. Fig. 2 shows how we represent $\mu$ individuals in form of texture. Without loss of generality, we take $k$=32 as an example of illustration throughout this paper.

As each pixel in the texture contains *quadruple* of 32-bit floating point values $(r, g, b, \alpha)$, we can encode an individual of 32 variables into 8 pixels. Instead of mapping an individual to 8 consecutive pixels in the texture, we divide an

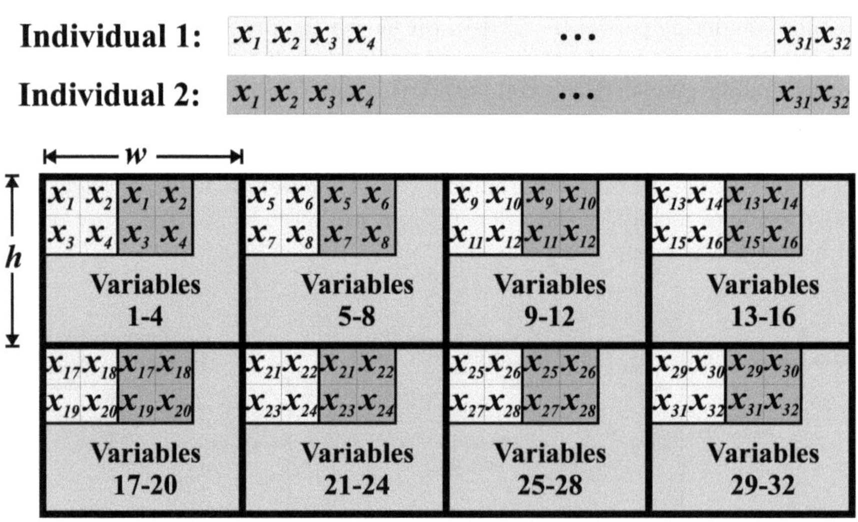


**Fig. 2.** Representing individuals of 32 variables on textures

individual into quadruple of 4 variables. The same quadruples from all individuals are grouped and form a tile in the texture as shown in Fig. 2. Each tile is $w \times h = \mu$ in size. The reason we do not adopt the *consecutive-pixel* representation is that the implementation will be complicated when $k$ varies. Imagine the complication of variables' offsets within the texture when $k$ increases from 32 to 48. On the other hand, the *fragmentation-and-tiling* representation is more scalable because increasing $k$ can be easily achieved by adding more tiles. In our specific example of $k = 32$, 4×2 tiles are formed. It is up to user to decide the organization of these tiles in the texture. The first tile (upper-left tile) in Fig. 2 stores variables 1 to 4, while the next tile stores variables 5 to 8, and so on.

Texture on GPU is not as flexible as main memory. Current GPUs impose several limitations. One of them is the size of texture must not exceed certain limit, *e.g.* 4096×4096 on nVidia GeforceFX 6800. In other words, to fit the whole population in one texture on our GPU, we must satisfy $k\mu \leq 4 \times 4096^2$. For extremely large populations with a large number of variables, multiple textures have to be used. Note that there are also limitation on the total number of textures that can be accessed simultaneously. The actual number varies on different GPU models. Normally, at least 16 textures can be supported.

## 5 Hybrid Genetic Algorithm on GPU

Without loss of generality, we assume the optimization is to minimize an objective function. Hence, our HGA is used to determine a $\mathbf{x}_{\mathrm{min}}$, such that

$$\forall \mathbf{x}, f(\mathbf{x}_{\mathrm{min}}) \leq f(\mathbf{x})$$

where $\mathbf{x} = \{x_i(1), x_i(2), \ldots, x_i(k)\}$ is the individual containing $k$ variables; $f : R^n \mapsto R$ is the function being optimized. The algorithm is given as follows:

1. Set $t = 0$.
2. Generate the initial population $P(t)$ of $\mu$ individuals, each of which can be represented as a set of real vectors, $(\mathbf{x_i}, \boldsymbol{\eta}_\mathbf{i}), i = 1, \ldots, \mu$. Both $\mathbf{x_i}$ and $\boldsymbol{\eta}_\mathbf{i}$ contain $k$ independent variables,
   $\mathbf{x_i} = \{x_i(1), \ldots, x_i(k)\}$
   $\boldsymbol{\eta}_\mathbf{i} = \{\eta_i(1), \ldots, \eta_i(k)\}$
3. Evaluate the fitness values of individuals in $P(t)$ by using a fitness measure based on the objective function to be optimized.
4. Return solution if termination condition is satisfied, else go to step 5.
5. For each individual $(\mathbf{x_i}, \boldsymbol{\eta}_\mathbf{i})$ in $P(t)$ denoted by $P_i^1$, where $i = 1, \ldots, \mu$, select two parents $P^1_{parent_i 1}$ and $P^1_{parent_i 2}$ from $P(t)$ using the tournament selection method.
6. For each $P_i^1$, recombine $P^1_{parent_i 1}$ and $P^1_{parent_i 2}$ using single point crossover to produce two offspring $P^2_{i1}$ and $P^2_{i2}$ that are stored in the temporary population $P^2$. The population $P^2$ contains $2\mu$ individuals.

7. Mutate individuals in $P^2$ to generate modified individuals that are stored in the temporary population $P^3$. For an individual $P^2_{il} = (\mathbf{x_{il}}, \boldsymbol{\eta}_{\mathbf{il}})$, where $i = 1, \ldots, \mu$ and $l = 1, 2$, create a new individual $P^3_{il} = (\mathbf{x_{il}}', \boldsymbol{\eta}_{\mathbf{il}}')$ as follows: for $j = 1, \ldots, k$
$x'_{il}(j) = x_{il}(j) + \eta_{il}(j)R(0,1)$,
$\eta'_{il}(j) = \eta_{il}(j)\exp(\frac{1}{\sqrt{2k}}R(0,1) + \frac{1}{\sqrt{2\sqrt{k}}}R_j(0,1))$
where $x_{il}(j), \eta_{il}(j), x'_{il}(j)$, and $\eta'_{il}(j)$ denote the $j$-th component of $\mathbf{x_{il}}, \boldsymbol{\eta}_{\mathbf{il}}, \mathbf{x_{il}}'$, and $\boldsymbol{\eta}_{\mathbf{il}}'$ respectively. $R(0,1)$ denotes a normally distributed 1D random number with zero mean and standard deviation of one. $R_j(0,1)$ indicates a new random value for each value of $j$.
8. Evaluate the fitness values of the individuals in population $P^3$.
9. For each individual $(\mathbf{x_i}, \boldsymbol{\eta}_{\mathbf{i}})$ in $P(t)$, compare $P^1_i$, $P^3_{i1}$ and $P^3_{i2}$, the one with the best fitness value will become an individual of the population $P(t+1)$ of the next generation.
10. $t = t + 1$.
11. Go to step 4.

In the above pseudocode, $\mathbf{x_i}$ is a vector of target variables evolving and $\boldsymbol{\eta}_{\mathbf{i}}$ controls the vigorousness of mutation of $\mathbf{x_i}$. In general, the computation of HGA can be roughly divided into four types: (a) fitness value evaluation (steps 3 and 8), (b) parent selection (step 5), (c) crossover and mutation (steps 6 and 7 respectively), and (d) the replacement schema designed for parallel algorithms (step 9). These types of operations will be discussed in the following sub-sections.

### 5.1 Fitness Value Evaluation

Fitness value evaluation determines the "goodness" of individuals. It is one of the core parts of HGA. After each evolution, the fitness value of each individual in the current population is calculated. The result is then passed to the later steps of HGA process. Each individual returns a fitness value by feeding the objective function $f$ with the target variables of the individual. This evaluation process usually consumes most of the computational time.

Since no interaction among individuals is required during evaluation, the evaluation is fully parallelizable. Recall that the individuals are broken down into quadruples and stored in the tiles within the textures. The evaluation shader hence looks up the corresponding quadruple in each tile during the evaluation. The fitness values are output to an output texture of size $w \times h$, instead of $4w \times 2h$, because each individual only returns a single value.

### 5.2 Parent Selection

The selection process determines which individuals will be selected as parents to reproduce offspring. The selection operators in genetic algorithms are not specific, however the fitness value of an individual usually induces a probability of being selected. The *roulette wheel selection, truncation selection*, and *stochastic tournament* are usually applied in genetic algorithms [3]. The stochastic tournament is employed in HGA because of the following reasons. Firstly, it is not practical to implement a parallel method on GPU to collect statistical information on the whole population. Since this information is not required in the stochastic tournament while it is needed for the other two methods, the stochastic tournament is more suitable for GPU. Secondly, it is generally believed that the tournament selection yields better performance for large populations. Thus, the tournament selection tends to be the mainstream selection strategy.

In the tournament selection method, two groups of $q$ individuals are randomly chosen from the population for each individual. The number $q$ is the tournament size. The two individuals with the best fitness values within the two groups will be selected as the parents to produce offspring by using crossover and mutation. The problem is how we sample the individuals from the population to form the tournament groups. This sampling is usually achieved by randomly choosing individuals from the whole population, and this is called the global selection method. According to [20], the tournament size $q$ for a small population ($\mu \leq 500$) should be set at about five percents to ten percents of the population size, i.e. for a population with 100 individuals, a minimum tournament size of 5 should be set. For larger populations, this criterion can be loosen. Since GPU lacks the power of generating random numbers, a large texture containing random numbers should be transferred from main memory to GPU memory first. The global selection process can then be performed in GPU. For the specific problems studied in this paper, $\mu \times q \times 2$ random numbers should be transferred to GPU memory. Due to the limit on the memory bandwidth, relatively long computation time is consumed to perform this transferral if the population size is large.

In the local selection method for the fine-grained parallel computation model, each individual can access all its local neighbors and select the best two individuals as the parents. This approach solves the previous random number transferring problem as it is a deterministic method which does not use any random numbers. However, this also imposes limitations that may lead to slow convergence. In this paper, We propose a new pseudo-deterministic selection method that reduces the number of random numbers being transferred while enhancing the

```
For ALL individual i
BEGIN
  randomly pick an address k
  parent_i = (address of individual i) + k
  For j = 1 to q − 1
  BEGIN
    randomly pick an address k
    if (fitness[parent_i + k] < fitness[parent_i])
      parent_i = parent_i + k
  END
END
```

**Fig. 3.** The pseudo-deterministic selection method

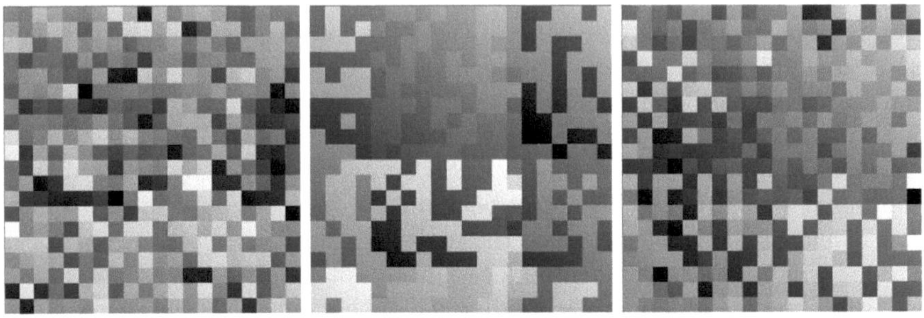

(a) (b) (c)

**Fig. 4.** Addresses of the selected parents. (a) Global selection, (b) Local selection, (c) Pseudo-deterministic selection.

exploitation of global information. The pseudo-deterministic selection method is described in Fig. 3.

The final addresses of parents for each individual are summarized in Fig. 4. We use the graph of the global selection approach (Fig. 4(a)) as a reference point, the noisy pattern illustrates that the resulting addresses are purely random. For the local selection method, the graph in Fig. 4(b) shows that neighboring individuals have the same colors. In other words, the individuals in the same local neighborhoods tend to select the same individuals as their parents. On the other hand, the graph for our pseudo-deterministic approach (Fig. 4(c)) illustrates a high degree of randomness, though there are still some tiny self-similar patterns. For the tournament size of $q$, the pseudo-deterministic approach requires to transfer only $q \times 2$ random numbers from main memory to GPU memory. On the other hand, $\mu \times q \times 2$ random numbers are transferred for the global selection approach. Thus, our approach is more efficient than the latter one. From our experiment results that will be discussed in Sub-section 6.3, we observe that the performance of our pseudo-deterministic approach is comparable to that of the global selection approach while the former is faster than the latter for large populations.

We implement our parent selection method in a fragment shader. The input of the shader is the texture containing the fitness values of the individuals, as well as $2 \times q$ random numbers. While the output of the shader is the addresses of the breeding parents selected. Recall that a pixel is composed of four 32-bit floating point values $(r, g, b, \alpha)$, the address of the first selected parent is stored in the $r$ component and the address of the second parent is stored in the $b$ component. Thus, the addresses of all selected parents are stored in an output texture of size $w \times h$.

### 5.3 Crossover and Mutation

The selection operator focuses on searching promising regions of the solution space. However, it is not able to introduce new solutions that are not in

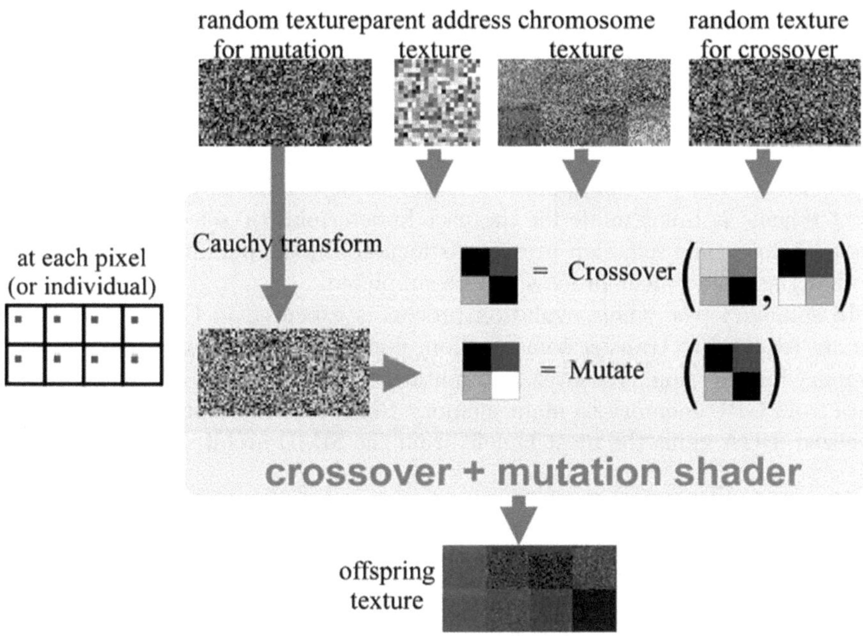


**Fig. 5.** Shader for performing recombination and mutation

the current population. In order to escape from local optima and introduce larger population diversity, the crossover and mutation operators are introduced. There are several ways to implement the crossover operator and we use *single point crossover* [3] in our genetic algorithm.

Since the crossover procedure uses only a small amount of computer resources, it is not efficient to devote a whole rendering pass to this operation. To optimize the performance, we implement the crossover and mutation operators in a fragment shader as depicted in Fig. 5. In this shader, the Cauchy mutation method proposed by Yao and Liu [14] is implemented. For the shader, four input textures are needed. The first one is the texture containing the individuals of the population. The second one is the texture containing addresses generated by the parent selection shader. The third one is the texture containing crossover positions of individuals. While the last one is a texture containing random numbers for the Cauchy mutation. In the shader program, each fragment acts as an tile of an individual, which grabs the corresponding parents tile according to the texture containing addresses of parents. The two selected parents will undergo the process of crossover and mutation to produce an offspring. Since we treat each fragment as an tile (containing four variables), and only one tile of an offspring can be generated per fragment. Thus the fragment shader is executed twice to produce two offspring for each individual. In the second execution, the roles of the first parent and the second parent are exchanged.

### 5.4 Replacement

Finally, we replace the population by comparing each individual with the two corresponding offspring. The one having the best fitness value replaces the individual. HGA continues to iterate until the predefined termination criterion is met. Unlike the sequential replacement scheme used in our parallel FEP, our HGA uses a fully parallel approach. It should be emphasized that this replacement scheme is tailor made for the pseudo-deterministic selection method discussed before. The selection pressure is focused on the parent selection process so that the replacement process can be simplified.

In summary, the whole evolution process is executed in GPU. Moreover, it is only required to transfer some random numbers from main memory to GPU memory for selection, crossover, and mutation, while it is not necessary to move data from GPU memory to main memory. Since the whole process is fully parallelized, HGA gains the most benefit from the SIMD architecture of GPU.

## 6 Experiment Results

We have compared HGA, FEP, GPU and CPU implementations of HGA, the pseudo-deterministic and the global selection methods on a set of benchmark optimization problems used in [15]. Table 1 summarizes the benchmark functions, the number of variables, the search ranges, and the minimum values of these functions. We have conducted the experiments for 20 trials and the average performance is reported in this paper. The experiment test bed was an AMD Athlon$^{TM}$ 64 3000+ CPU with a PCI Express enabled consumer-level GeForce 6800 Ultra display card, with 1,024 MB main memory and 256 MB GPU memory. The following parameters have been used in the experiments:

- population size: $\mu = 400, 800, 3200, 6400$
- tournament size: $q = 10$
- maximum number of generation: $G = 2000$

### 6.1 Comparison between HGA and FEP

We have compared the performance of our GPU implementations of HGA and FEP on the functions listed in Table 1. Fig. 6 and Fig. 7 depict, by generation, the average fitness value of the best solutions found by HGA and FEP with different population sizes in 20 trials. It can be observed that better solutions can be obtained for all functions if a larger population size is used. Moreover, HGA finds better solutions and converges much faster than FEP for all functions and population sizes. This phenomenon demonstrates the effectiveness of the crossover operator, which recombines good building blocks to generate better solutions, and thus results in faster convergence.

The average execution time of HGA and FEP for different population sizes is summarized in Table 2. It can be found that HGA executes much faster than FEP for all functions and population sizes. For HGA, fitness value evaluation,

**Table 1.** The set of test functions. The number of variables $N = 32$ for all test functions. $S$ indicates the ranges of the variables and $f_m$ is the minimum value of the function.

| Test Functions | S | $f_m$ |
|---|---|---|
| $f_1 : \sum_{i=1}^{N} x_i^2$ | $(-100, 100)^N$ | 0 |
| $f_2 : \sum_{i=1}^{N} (\sum_{j=1}^{i} x_j)^2$ | $(-100, 100)^N$ | 0 |
| $f_3 : \sum_{i=1}^{N-1} \{100(x_{i+1} - x_i^2)^2 + (x_i - 1)^2\}$ | $(-30, 30)^N$ | 0 |
| $f_4 : -\sum_{i=1}^{N} x_i \sin(\sqrt{\lvert x_i \rvert})$ | $(-500, 500)^N$ | -13407.36 |
| $f_5 : \sum_{i=1}^{N} \{x_i^2 - 10\cos(2\pi x_i) + 10\}$ | $(-5.12, 5.12)^N$ | 0 |
| $f_6 : -20exp\{-0.2\sqrt{\frac{1}{N}\sum_{i=1}^{N} x_i^2}\} - exp\{\frac{1}{N}\sum_{i=1}^{N} cos(2\pi x_i)\} + 20 + e$ | $(-32, 32)^N$ | 0 |
| $f_7 : \frac{1}{4000}\sum_{i=1}^{N} x_i^2 - \prod_{i=1}^{N} cos(\frac{x_i}{\sqrt{i}}) + 1$ | $(-600, 600)^N$ | 0 |
| $f_8 : \sum_{i=1}^{N} \lvert x_i \rvert + \prod_{i=1}^{N} \lvert x_i \rvert$ | $(-10, 10)^N$ | 0 |
| $f_9 : \sum_{i=1}^{N} \lfloor x_i + 0.5 \rfloor$ | $(-100, 100)^N$ | 0 |

**Table 2.** The average execution time (in seconds) of HGA and FEP with different population sizes

| | HGA | | | | | | | | | FEP | | | | | | | | |
|---|---|---|---|---|---|---|---|---|---|---|---|---|---|---|---|---|---|---|
| $\mu$ | $f_1$ | $f_2$ | $f_3$ | $f_4$ | $f_5$ | $f_6$ | $f_7$ | $f_8$ | $f_9$ | $f_1$ | $f_2$ | $f_3$ | $f_4$ | $f_5$ | $f_6$ | $f_7$ | $f_8$ | $f_9$ |
| 400 | 6.03 | 6.04 | 6.17 | 6.03 | 6.13 | 6.40 | 6.24 | 6.18 | 6.13 | 33.66 | 33.64 | 33.73 | 33.64 | 33.70 | 33.69 | 33.78 | 33.62 | 33.66 |
| 800 | 7.99 | 8.03 | 8.15 | 8.04 | 8.13 | 8.13 | 8.22 | 8.05 | 8.13 | 33.66 | 33.64 | 33.72 | 33.64 | 33.68 | 33.69 | 33.78 | 33.62 | 33.67 |
| 3200 | 19.8 | 19.9 | 19.9 | 20.3 | 19.9 | 20.02 | 20.09 | 19.88 | 19.93 | 34.22 | 34.60 | 34.69 | 34.60 | 34.75 | 33.76 | 33.86 | 33.70 | 33.73 |
| 6400 | 36.1 | 36.1 | 36.4 | 37.7 | 36.4 | 36.36 | 36.37 | 36.52 | 36.32 | 67.46 | 67.26 | 67.31 | 67.23 | 68.47 | 67.16 | 67.25 | 67.10 | 67.14 |

parent selection, crossover and mutation, and replacement are all executed in GPU. The CPU only generates a number of random numbers and passes them to GPU through some input textures. On the other hand, fitness value evaluation, mutation, and reproduction of FEP are executed in GPU. Selection, replacement, and random number generation are performed in CPU. Consequently, our GPU implementation of HGA is much more efficient than our GPU implementation of FEP.

### 6.2 Comparison between GPU and CPU Approaches

We have performed experiments to compare our GPU and CPU implementations of HGA. From the experiment results displayed in Fig. 6 and Fig. 7, we find that better solutions can be obtained by HGA for all functions if a larger population size is used. However, HGA with a larger population size will take longer execution time. Fig. 8 and Fig. 9 display, by generation, the average execution time of the GPU and CPU implementations of HGA with different population sizes. From the curves in these figures, the execution time increases if a larger population is applied. However, our GPU approach is much more efficient than

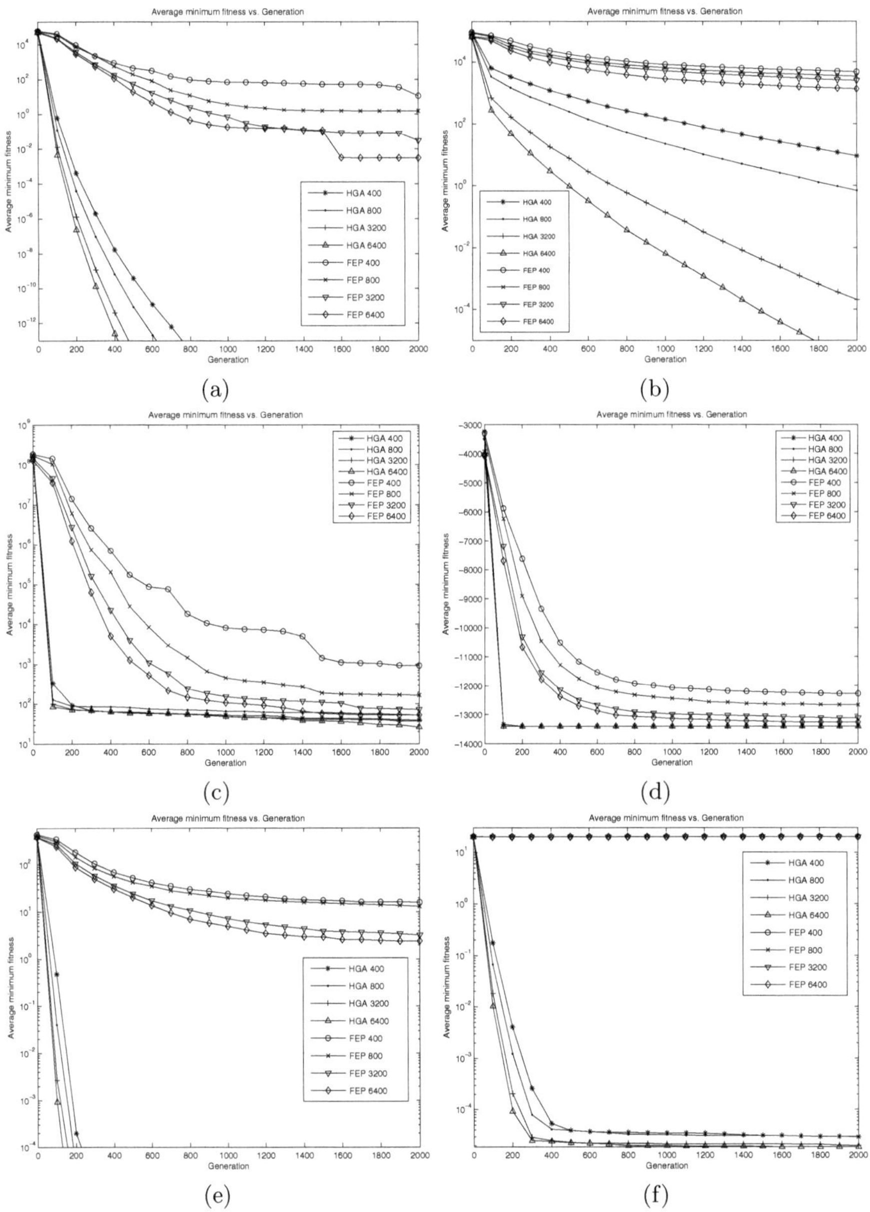


**Fig. 6.** Fitness value of the best solution found by HGA and FEP for functions $f_1$ - $f_6$. The results were averaged over 20 independent trials. (a)-(f) correspond to functions $f_1$ - $f_6$ respectively.

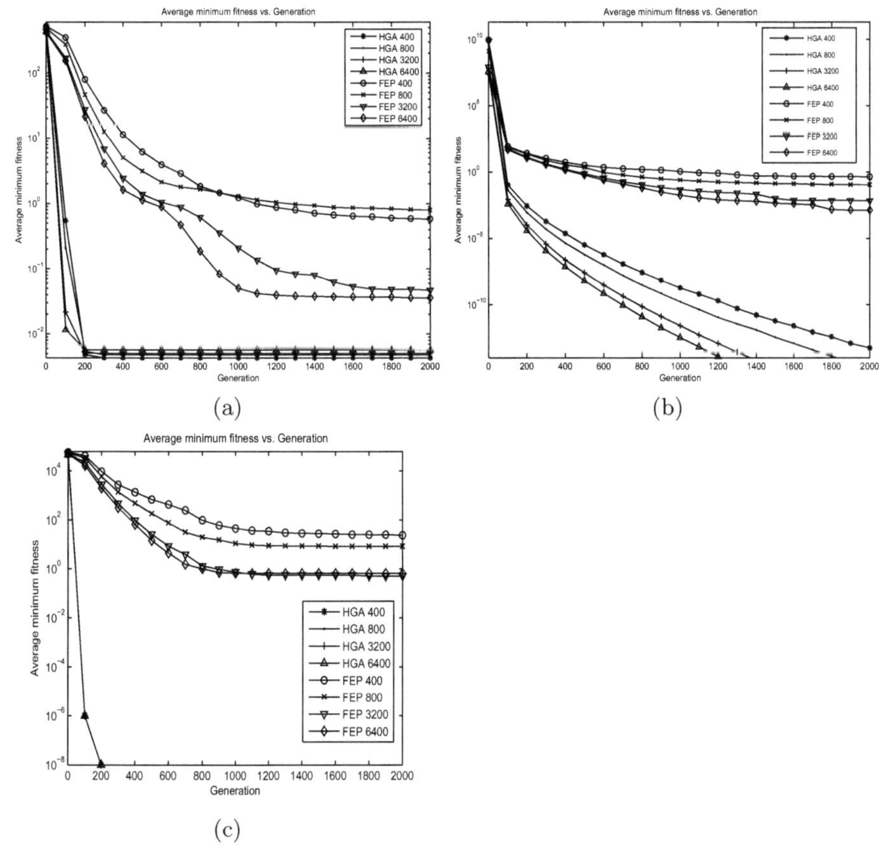


**Fig. 7.** Fitness value of the best solution found by HGA and FEP for functions $f_7$ - $f_9$. The results were averaged over 20 independent trials. (a)-(c) correspond to functions $f_7$ - $f_9$ respectively.

the CPU implementation because the execution time of the former is much less than that of the latter. Moreover, the efficiency leap becomes larger when the population size increases.

The ratios of the average execution time of the GPU (CPU) approach with population sizes of 800, 3200, and 6400 to that of the corresponding approach with population size of 400 are summarized in Table 3. It is interesting to notice that, the CPU approach shows a linear relation between the number of individuals and the execution time, while our GPU approach has a sub-linear relation. For example, the execution time of the GPU approach with population size of 6400 is about 6 times of that with population size of 400. Definitely, this is an advantage when huge population sizes are required in some real-life applications.

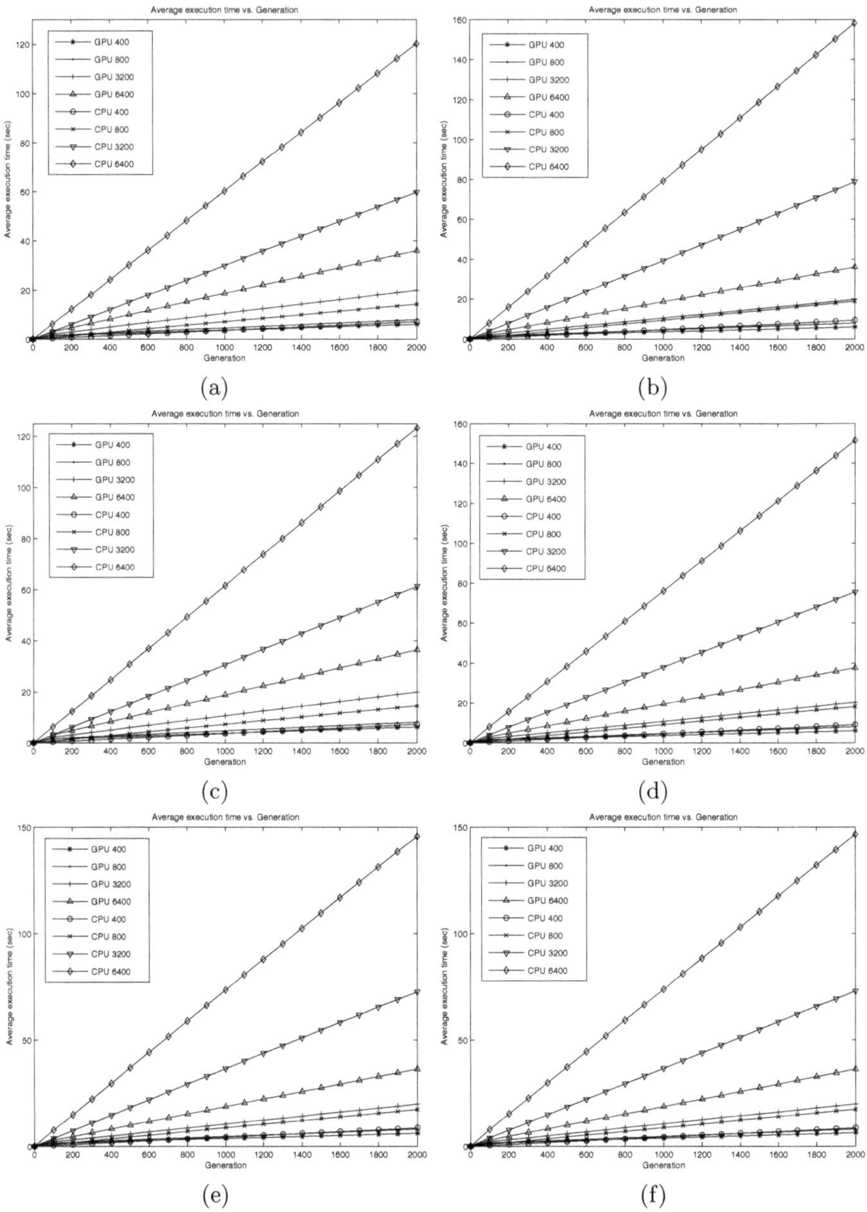


**Fig. 8.** The average execution time of the GPU and CPU approaches for functions $f_1$ - $f_6$. The results were averaged over 20 independent trials. (a)-(f) correspond to functions $f_1$ - $f_6$ respectively.

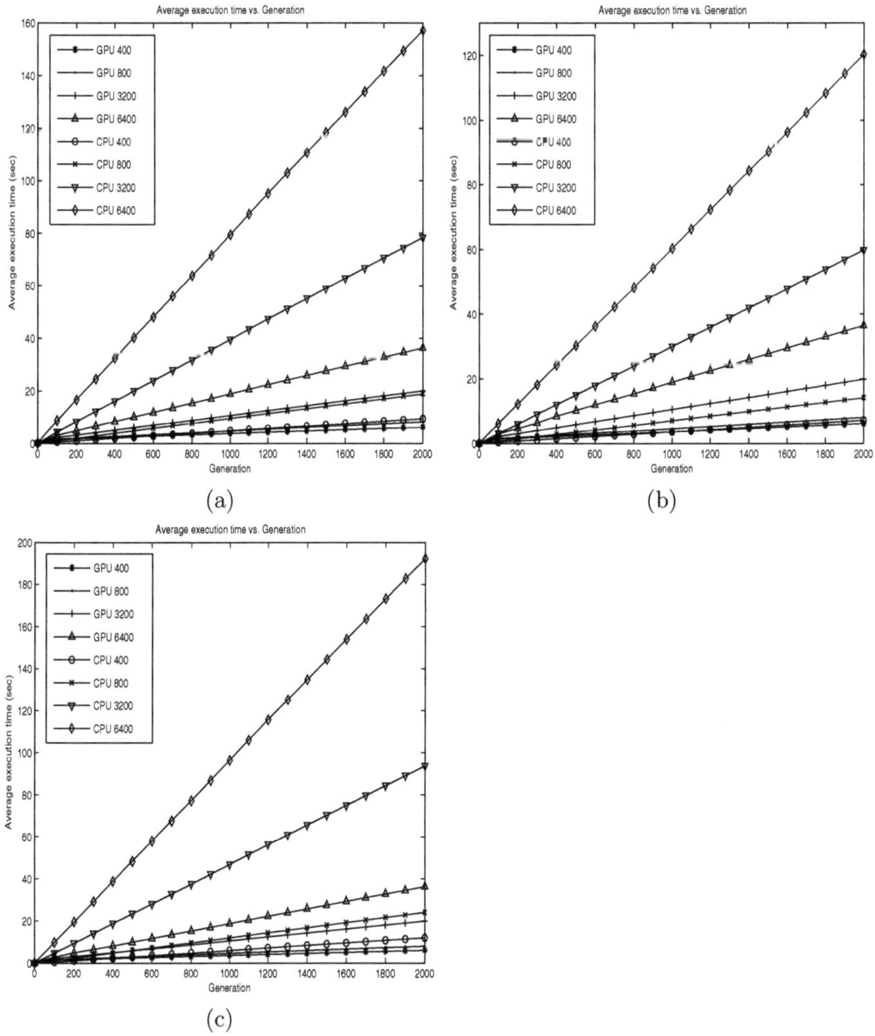


**Fig. 9.** The average execution time of the GPU and CPU approaches for functions $f_7$ - $f_9$. The results were averaged over 20 independent trials. (a)-(c) correspond to functions $f_7$ - $f_9$ respectively.

Table 4 displays the speed-ups of our GPU approach with the CPU approach. The speed-ups depend on the population size and the problem complexity. They range from about 1.16 to about 5.30. For complicated problems that require huge population sizes, we expect that GPU can achieve even better performance gain.

**Table 3.** The ratios of the average execution time of the GPU (CPU) approach with different population sizes to that with population size of 400

| $\mu$ | GPU | | | | | | | | | CPU | | | | | | | | |
|---|---|---|---|---|---|---|---|---|---|---|---|---|---|---|---|---|---|---|
| | $f_1$ | $f_2$ | $f_3$ | $f_4$ | $f_5$ | $f_6$ | $f_7$ | $f_8$ | $f_9$ | $f_1$ | $f_2$ | $f_3$ | $f_4$ | $f_5$ | $f_6$ | $f_7$ | $f_8$ | $f_9$ |
| 800 | 1.33 | 1.33 | 1.32 | 1.33 | 1.33 | 1.27 | 1.32 | 1.30 | 1.33 | 2.00 | 2.00 | 2.00 | 2.00 | 2.00 | 2.00 | 2.00 | 1.97 | 2.00 |
| 3200 | 3.28 | 3.29 | 3.24 | 3.36 | 3.25 | 3.13 | 3.22 | 3.21 | 3.25 | 8.46 | 8.34 | 8.46 | 8.33 | 8.34 | 8.34 | 8.29 | 8.33 | 7.82 |
| 6400 | 5.99 | 5.98 | 5.91 | 6.25 | 5.94 | 5.68 | 5.83 | 5.90 | 5.92 | 17.01 | 16.75 | 17.03 | 16.69 | 16.72 | 16.76 | 16.63 | 16.74 | 16.04 |

**Table 4.** The speed-up of the GPU approach

| $\mu$ | $f_1$ | $f_2$ | $f_3$ | $f_4$ | $f_5$ | $f_6$ | $f_7$ | $f_8$ | $f_9$ |
|---|---|---|---|---|---|---|---|---|---|
| 400 | 1.17 | 1.56 | 1.17 | 1.50 | 1.42 | 1.37 | 1.51 | 1.16 | 1.96 |
| 800 | 1.77 | 2.35 | 1.78 | 2.26 | 2.14 | 2.15 | 2.30 | 1.76 | 2.95 |
| 3200 | 3.02 | 3.96 | 3.07 | 3.73 | 3.64 | 3.64 | 3.90 | 3.01 | 4.71 |
| 6400 | 3.33 | 4.38 | 3.38 | 4.02 | 4.00 | 4.03 | 4.32 | 3.30 | 5.30 |

### 6.3 Comparison between Global and Pseudo-deterministic Selections

To study the effectiveness and efficiency of our pseudo-deterministic selection method, we have compared two different GPU implementations of HGA. The first one uses our pseudo-deterministic selection method while the other employs the global selection approach. For the 20 trials of the experiments for different functions and population sizes, the best fitness values achieved by the two implementations in each generation have been recorded. These fitness values have been analysed by using a two-tailed $t$-test with a significance level of 0.05. It has been found that there is no significant difference between the fitness values obtained by the two implementations for all functions and population sizes. In other words, our pseudo-deterministic selection method achieves similar performance as the global selection method.

The speed-ups of our pseudo-deterministic selection method with the global selection approach are summarized in Table 5. They range from about 0.92 to about 1.22. It can be observed that our selection method improves the execution time of HGA when the population size is greater than or equal to 800. Moreover, our selection method is more efficient for larger population size.

**Table 5.** The speed-up of Pseudo-deterministic selection

| $\mu$ | $f_1$ | $f_2$ | $f_3$ | $f_4$ | $f_5$ | $f_6$ | $f_7$ | $f_8$ | $f_9$ |
|---|---|---|---|---|---|---|---|---|---|
| 400 | 1.02 | 0.95 | 0.95 | 0.95 | 0.95 | 0.92 | 0.95 | 0.97 | 0.95 |
| 800 | 1.08 | 1.01 | 1.01 | 1.01 | 1.01 | 0.99 | 1.01 | 1.01 | 1.01 |
| 3200 | 1.22 | 1.16 | 1.16 | 1.13 | 1.16 | 1.15 | 1.15 | 1.16 | 1.19 |
| 6400 | 1.22 | 1.19 | 1.19 | 1.14 | 1.18 | 1.18 | 1.18 | 1.18 | 1.19 |

## 7 Conclusion

In this research, we have implemented a parallel HGA on GPU which are available and installed on ubiquitous personal computers. HGA extends the classical genetic algorithm by incorporating the Cauchy mutation operator from evolutionary programming. In our parallel HGA, all steps except the random number generation procedure are performed in GPU and thus our parallel HGA can be executed efficiently. We have proposed the pseudo-deterministic selection method which is comparable to the global selection approach with significant execution time performance advantages.

We have done experiments to compare our parallel HGA and our previous parallel FEP. It is found that HGA converges much faster than FEP for all test functions and population sizes. Moreover, the average execution time of HGA is much smaller than that of FEP for all test functions and population sizes. In other words, our parallel HGA is more effective and efficient than our previous parallel FEP. We have performed experiments to compare our parallel HGA and a CPU implementation of HGA. It is found that the speed-up factor of our parallel HGA ranges from 1.16 to 5.30. Moreover, there is a sub-linear relation between the population size and the execution time. Thus, our parallel HGA will be very useful for solving difficult problems that require huge population sizes. Our pseudo-deterministic selection method has been examined and it is found that this method is effective and efficient for our parallel HGA.

With the wide availability of GPU, GA running GPU can benefit people in various applications requiring fast optimized solution. Potential applications includes artificial intelligence engine in computer games, and fast time-table scheduling, etc.

For future work, we plan to compare our parallel HGA and parallel FEP on more test functions and study the effects of different parameters (such as the population size and the tournament size) on the performance of these parallel algorithms. We are also parallelizing other kinds of evolutionary algorithms including genetic programming [21, 22, 23] and evolution strategies [24, 25].

## Acknowledgment

This work is supported by the Lingnan University Direct Grant DR08B2.

## References

1. Holland, J.H.: Adaptation in Natural and Artificial Systems. University of Michigan Press (1975)
2. Oh, I.S., Lee, J.S., Moon, B.R.: Hybrid Genetic Algorithms for Feature Selection. IEEE Transactions on Pattern Analysis and Machine Intelligence 26(11), 1424–1437 (2004)
3. Goldberg, D.E.: Genetic Algorithms in Search, Optimization, and Machine Learning. Addison-Wesley, Reading (1989)
4. Freitas, A.A.: Data Mining and Knowledge Discovery with Evolutionary Algorithms. Springer, Heidelberg (2002)
5. Myers, J.W., Laskey, K.B., DeJong, K.A.: Learning Bayesian Networks from Incomplete Data using Evolutionary Algorithms. In: Proceedings of the First Annual Conference on Genetic and Evolutionary Computation Conference, pp. 458–465 (1999)

6. Larrañaga, P., Poza, M., Yurramendi, Y., Murga, R., Kuijpers, C.: Structural Learning of Bayesian Network by Genetic Algorithms: A Performance Analysis of Control Parameters. IEEE Transactions on Pattern Analysis and Machine Intelligence 18(9), 912–926 (1996)
7. GPGPU: General-Purpose Computation Using Graphics Hardware, http://www.gpgpu.org/
8. Moreland, K., Angel, E.: The FFT on a GPU. In: Proceedings of 2003 SIGGRAPH/Eurographics Workshop on Graphics Hardware, pp. 112–119 (2003)
9. Wang, J.Q., Wong, T.T., Heng, P.A., Leung, C.S.: Discrete Wavelet Transform on GPU. In: Proceedings of ACM Workshop on General Purpose Computing on Graphics Processors C-41 (2004)
10. Jiang, C., Snir, M.: Automatic Tuning Matrix Multiplication Performance on Graphics Hardware. In: Proceedings of the 14th International Conference on Parallel Architectures and Compilation Techniques, pp. 185–196 (2005)
11. Galoppo, N., Govindaraju, N.K., Henson, M., Manocha, D.: LU-GPU: Efficient Algorithms for Solving Dense Linear Systems on Graphics Hardware. In: Proceedings of the ACM/IEEE SC 2005 Conference 3 (2005)
12. Fok, K.L., Wong, T.T., Wong, M.L.: Evolutionary Computing on Consumer-Level Graphics Hardware. IEEE Intelligent Systems 22(2), 69–78 (2007)
13. Wong, M.L., Wong, T.T., Fok, K.L.: Parallel Evolutionary Algorithms on Graphics Processing Unit. In: Proceedings of IEEE Congress on Evolutionary Computation 2005 (CEC 2005), pp. 2286–2293 (2005)
14. Yao, X., Liu, Y.: Fast Evolutionary Programming. In: Proceedings of the 5th Annual Conference on Evolutionary Programming, pp. 451–460 (1996)
15. Yao, X., Liu, Y., Lin, G.: Evolutionary Programming Made Faster. IEEE Transactions on Evolutionary Computation 3(2), 82–102 (1999)
16. Fogel, D.B.: Evolutionary Computation: Toward a New Philosohpy of Machine Intelligence. IEEE Press, Los Alamitos (2000)
17. Fogel, L., Owens, A., Walsh, M.: Artificial Intelligence Through Simulated Evolution. John Wiley and Sons, Chichester (1966)
18. Angeline, P.: Genetic Programming and Emergent Intelligent. In: Kinnear, K.E. (ed.) Advances in Genetic Programming, pp. 75–97. MIT Press, Cambridge (1994)
19. Cantú-Paz, E.: Efficient and Accurate Parallel Genetic Algorithms. Kluwer Academic Publishers, Dordrecht (2000)
20. Bäck, T., Fogel, D.B., Michalewicz, Z.: Evolutionary Computation 2: Advanced Algorithms and Operators. Insitute of Physic Publishing (2000)
21. Koza, J.R.: Genetic Programming: On the Programming of Computers by Means of Natural Selection. MIT Press, Cambridge (1992)
22. Koza, J.R., Keane, M.A., Streeter, M.J., Mydlowec, W., Yu, J., Lanza, G.: Genetic Programming IV: Routine Human-Competitive Machine Intelligence. Kluwer Academic Publishers, Dordrecht (2003)
23. Banzhaf, W., Nordin, P., Keller, R.E., Francone, F.D.: Genetic Programming: An Introduction. Morgan Kaufmann, San Francisco (1998)
24. Schewefel, H.P.: Numerical Optimization of Computer Models. John Wiley and Sons, Chichester (1981)
25. Bäck, T.: Evolutionary Algorithms in Theory and Practice: Evolution Strategies, Evolutionary Programming, Genetic Algorithms. Oxford University Press, Oxford (1996)

# Author Index

MIX
Papier aus verantwortungsvollen Quellen
Paper from responsible sources
FSC® C105338

If you have any concerns about our products,
you can contact us on
**ProductSafety@springernature.com**

In case Publisher is established outside the EU,
the EU authorized representative is:
**Springer Nature Customer Service Center GmbH**
**Europaplatz 3, 69115 Heidelberg, Germany**

Printed by Libri Plureos GmbH
in Hamburg, Germany